GACE Mathematics

022 Teacher Certification Exam
023

By: Sharon Wynne, M.S
Southern Connecticut State University

XAMonline, INC.
Boston

To obtain permission(s) to use the material from this work for any purpose including workshops or seminars, please submit a written request to:

XAMonline, Inc.
21 Orient Ave.
Melrose, MA 02176
Toll Free 1-800-509-4128
Email: info@xamonline.com
Web www.xamonline.com
Fax: 1-781-662-9268

Library of Congress Cataloging-in-Publication Data

Wynne, Sharon A.
  GACE: Mathematics 022, 023 Teacher Certification / Sharon A. Wynne. -2nd ed.
  ISBN: 978-1-58197-346-4
  1. Mathematics 022, 023   2. Study Guides.   3. GACE
  4. Teachers' Certification & Licensure.   5. Careers

Disclaimer:
The opinions expressed in this publication are the sole works of XAMonline and were created independently from the National Education Association, Educational Testing Service, or any State Department of Education, National Evaluation Systems or other testing affiliates.

Between the time of publication and printing, state specific standards as well as testing formats and website information may change that is not included in part or in whole within this product. Sample test questions are developed by XAMonline and reflect similar content as on real tests; however, they are not former tests. XAMonline assembles content that aligns with state standards but makes no claims nor guarantees teacher candidates a passing score. Numerical scores are determined by testing companies such as NES or ETS and then are compared with individual state standards. A passing score varies from state to state.

Printed in the United States of America   œ-1

GACE: Mathematics 022, 023
ISBN: 978-1-58197-346-4

Table of Contents

SUBAREA VI.　　　MATHEMATICAL PROCESSES AND PERSPECTIVES

Great Study and Testing Tips!

What to study in order to prepare for the subject assessments is the focus of this study guide but equally important is *how* you study.

You can increase your chances of truly mastering the information by taking some simple, but effective steps.

Study Tips:

1. <u>Some foods aid the learning process</u>. Foods such as milk, nuts, seeds, rice, and oats help your study efforts by releasing natural memory enhancers called CCKs (*cholecystokinin*) composed of *tryptophan*, *choline*, and *phenylalanine*. All of these chemicals enhance the neurotransmitters associated with memory. Before studying, try a light, protein-rich meal of eggs, turkey, and fish. All of these foods release the memory enhancing chemicals. The better the connections, the more you comprehend.

Likewise, before you take a test, stick to a light snack of energy boosting and relaxing foods. A glass of milk, a piece of fruit, or some peanuts all release various memory-boosting chemicals and help you to relax and focus on the subject at hand.

2. <u>Learn to take great notes</u>. A by-product of our modern culture is that we have grown accustomed to getting our information in short doses (i.e. TV news sound bites or USA Today style newspaper articles.)

Consequently, we've subconsciously trained ourselves to assimilate information better in <u>neat little packages</u>. If your notes are scrawled all over the paper, it fragments the flow of the information. Strive for clarity. Newspapers use a standard format to achieve clarity. Your notes can be much clearer through use of proper formatting. A very effective format is called the *"Cornell Method."*

> Take a sheet of loose-leaf lined notebook paper and draw a line all the way down the paper about 1-2" from the left-hand edge.
>
> Draw another line across the width of the paper about 1-2" up from the bottom. Repeat this process on the reverse side of the page.

Look at the highly effective result. You have ample room for notes, a left hand margin for special emphasis items or inserting supplementary data from the textbook, a large area at the bottom for a brief summary, and a little rectangular space for just about anything you want.

3. <u>Get the concept then the details</u>. Too often we focus on the details and don't gather an understanding of the concept. However, if you simply memorize only dates, places, or names, you may well miss the whole point of the subject.

A key way to understand things is to put them in your own words. If you are working from a textbook, automatically summarize each paragraph in your mind. If you are outlining text, don't simply copy the author's words.

Rephrase them in your own words. You remember your own thoughts and words much better than someone else's, and subconsciously tend to associate the important details to the core concepts.

4. <u>Ask Why?</u> Pull apart written material paragraph by paragraph and don't forget the captions under the illustrations.

Example: If the heading is "Stream Erosion", flip it around to read "Why do streams erode?" Then answer the questions.

If you train your mind to think in a series of questions and answers, not only will you learn more, but it also helps to lessen the test anxiety because you are used to answering questions.

5. <u>Read for reinforcement and future needs</u>. Even if you only have 10 minutes, put your notes or a book in your hand. Your mind is similar to a computer; you have to input data in order to have it processed. *By reading, you are creating the neural connections for future retrieval.* The more times you read something, the more you reinforce the learning of ideas.

Even if you don't fully understand something on the first pass, *your mind stores much of the material for later recall.*

6. <u>Relax to learn so go into exile</u>. Our bodies respond to an inner clock called biorhythms. Burning the midnight oil works well for some people, but not everyone.

If possible, set aside a particular place to study that is free of distractions. Shut off the television, cell phone, and pager and exile your friends and family during your study period.

If you really are bothered by silence, try background music. Light classical music at a low volume has been shown to aid in concentration over other types. Music that evokes pleasant emotions without lyrics is highly suggested. Try just about anything by Mozart. It relaxes you.

7. <u>**Use arrows not highlighters.**</u> At best, it's difficult to read a page full of yellow, pink, blue, and green streaks. Try staring at a neon sign for a while and you'll soon see that the horde of colors obscure the message.

A quick note, a brief dash of color, an underline, and an arrow pointing to a particular passage is much clearer than a horde of highlighted words.

8. <u>**Budget your study time.**</u> Although you shouldn't ignore any of the material, *allocate your available study time in the same ratio that topics may appear on the test.*

Testing Tips:

1. Get smart, play dumb. Don't read anything into the question. Don't make an assumption that the test writer is looking for something else than what is asked. Stick to the question as written and don't read extra things into it.

2. Read the question and all the choices *twice* before answering the question. You may miss something by not carefully reading, and then re-reading both the question and the answers.

If you really don't have a clue as to the right answer, leave it blank on the first time through. Go on to the other questions, as they may provide a clue as to how to answer the skipped questions.

If later on, you still can't answer the skipped ones . . . *Guess.* The only penalty for guessing is that you *might* get it wrong. Only one thing is certain; if you don't put anything down, you will get it wrong!

3. Turn the question into a statement. Look at the way the questions are worded. The syntax of the question usually provides a clue. Does it seem more familiar as a statement rather than as a question? Does it sound strange?

By turning a question into a statement, you may be able to spot if an answer sounds right, and it may also trigger memories of material you have read.

4. Look for hidden clues. It's actually very difficult to compose multiple-foil (choice) questions without giving away part of the answer in the options presented.

In most multiple-choice questions you can often readily eliminate one or two of the potential answers. This leaves you with only two real possibilities and automatically your odds go to Fifty-Fifty for very little work.

5. Trust your instincts. For every fact that you have read, you subconsciously retain something of that knowledge. On questions that you aren't really certain about, go with your basic instincts. **Your first impression on how to answer a question is usually correct.**

6. Mark your answers directly on the test booklet. Don't bother trying to fill in the optical scan sheet on the first pass through the test.

Just be very careful not to miss-mark your answers when you eventually transcribe them to the scan sheet.

7. Watch the clock! You have a set amount of time to answer the questions. Don't get bogged down trying to answer a single question at the expense of 10 questions you can more readily answer.

SUBAREA I.   NUMBER CONCEPTS AND OPERATIONS

COMPETENCY 0001   UNDERSTAND NUMBER OPERATIONS AND BASIC PRINCIPLES OF NUMBER   THEORY.

The **absolute value** of a real number is the positive value of that number.

$|x| = x$ when $x \geq 0$ and

$|x| = x$ when $x < 0$.

Example: $|y - 7| = 2$

$y - 7 = 2$ or $y - 7 = -2$
$y = 9$ or $y = 5$

The solutions must be checked.

$|y - 7| = 2$ $|y - 7| = 2$
$|9 - 7|$? 2 $|5 - 7|$? 2
$|2|$? 2 $|-2|$? 2
2 = 2   2 = 2
true   true

Identify elements and subsets of the real number system.

a. **Natural numbers**--the counting numbers, 1,2,3,...

b. **Whole numbers**--the counting numbers along with zero, 0,1,2...

c. **Integers**--the counting numbers, their opposites, and zero, ..., ⁻1,0,1,...

d. **Rationals**--all of the fractions that can be formed from the whole numbers. Zero cannot be the denominator. In decimal form, these numbers will either be terminating or repeating decimals. Simplify square roots to determine if the number can be written as a fraction.

e. **Irrationals**--real numbers that cannot be written as a fraction. The decimal forms of these numbers are neither terminating nor repeating. Examples: $\pi, e, \sqrt{2}$, etc.

f. **Real numbers**--the set of numbers obtained by combining the rationals and irrationals. Complex numbers, i.e. numbers that involve i or $\sqrt{-1}$, are not real numbers.

Compare the relative size of real numbers expressed in a variety of forms, including fractions, decimals, percents, and scientific notation.

To convert a fraction to a decimal, simply divide the numerator (top) by the denominator (bottom). Use long division if necessary.

If a decimal has a fixed number of digits, the decimal is said to be terminating. To write such a decimal as a fraction, first determine what place value the farthest right digit is in, for example: tenths, hundredths, thousandths, ten thousandths, hundred thousands, etc. Then drop the decimal and place the string of digits over the number given by the place value.

If a decimal continues forever by repeating a string of digits, the decimal is said to be repeating. To write a repeating decimal as a fraction, follow these steps.

a. Let $x =$ the repeating decimal
    (ex. $x = .716716716...$)
b. Multiply x by the multiple of ten that will move the decimal just to the right of the repeating block of digits.
    (ex. $1000x = 716.716716...$)
c. Subtract the first equation from the second.
    (ex. $1000x - x = 716.716.716... - .716716...$)
d. Simplify and solve this equation. The repeating block of digits will subtract out.
    (ex. $999x = 716$ so $x = {}^{716}\!/_{999}$)
e. The solution will be the fraction for the repeating decimal.

Identify the greatest common factor (GCF) and least common multiple (LCM) of sets of numbers.

GCF is the abbreviation for the **greatest common factor**. The GCF is the largest number that is a factor of all the numbers given in a problem. The GCF can be no larger than the smallest number given in the problem. If no other number is a common factor, then the GCF will be the number 1. To find the GCF, list all possible factors of the smallest number given (include the number itself). Starting with the largest factor (which is the number itself), determine if it is also a factor of all the other given numbers. If so, that is the GCF. If that factor doesn't work, try the same method on the next smaller factor. Continue until a common factor is found. That is the GCF. Note: There can be other common factors besides the GCF.

Example: Find the GCF of 12, 20, and 36.

The smallest number in the problem is 12. The factors of 12 are 1,2,3,4,6 and 12. 12 is the largest factor, but it does not divide evenly into 20. Neither does 6, but 4 will divide into both 20 and 36 evenly.
Therefore, 4 is the GCF.

Example: Find the GCF of 14 and 15.

Factors of 14 are 1,2,7 and 14. 14 is the largest factor, but it does not divide evenly into 15. Neither does 7 or 2. Therefore, the only factor common to both 14 and 15 is the number 1, the GCF.

LCM is the abbreviation for **least common multiple**. The least common multiple of a group of numbers is the smallest number that all of the given numbers will divide into. The least common multiple will always be the largest of the given numbers or a multiple of the largest number.

Example: Find the LCM of 20, 30 and 40.

The largest number given is 40, but 30 will not divide evenly into 40. The next multiple of 40 is 80 (2 x 40), but 30 will not divide evenly into 80 either. The next multiple of 40 is 120. 120 is divisible by both 20 and 30, so 120 is the LCM (least common multiple).

- The Fundamental Theorem of Arithmetic states that every composite (non-prime) number can be written as a product of primes in one, and only one way.

  Prime numbers are whole numbers greater than 1 that have only 2 factors, 1 and the number itself. Examples of prime numbers are 2,3,5,7,11,13,17, or 19. Note that 2 is the only even prime number. When factoring into prime factors, all the factors must be numbers that cannot be factored again (without using 1). Initially numbers can be factored into any 2 factors. Check each resulting factor to see if it can be factored again. Continue factoring until all remaining factors are prime. This is the list of prime factors. Regardless of what way the original number was factored, the final list of prime factors will always be the same.

Remember that the number 1 is neither prime nor composite.

Example: Factor 30 into prime factors.

Factor 30 into any 2 factors.
$5 \cdot 6$ Now factor the 6.
$5 \cdot 2 \cdot 3$ These are all prime factors.
Factor 30   Into any 2 factors.
$3 \cdot 10$ Now factor the 10.
$3 \cdot 2 \cdot 5$ These are the same prime factors even though the original factors were different.

Example: Factor 240 into prime factors.

Factor 240   into any 2 factors.
$24 \cdot 10$ Now factor both 24 and 10.
$4 \cdot 6 \cdot 2 \cdot 5$ Now factor both 4 and 6.
$2 \cdot 2 \cdot 2 \cdot 3 \cdot 2 \cdot 5$ These are prime factors.

This can also be written as $2^4 \cdot 3 \cdot 5$.

COMPETENCY 0002   UNDERSTAND THE REAL AND COMPLEX NUMBER SYSTEMS.

Real numbers exhibit the following addition and multiplication **properties**, where a, b, and c are real numbers.

Note: Multiplication is implied when there is no symbol between two variables. Thus, $a \times b$ can be written ab. Multiplication can also be indicated by a raised dot ·

Closure
For all real numbers a and b,
$a + b$ is a unique real number.
ab is a unique real number.

Example: Since 2 and 5 are both real numbers, 7 is also a real number.

Example: Since 3 and 4 are both real numbers, 12 is also a real number.

Commutative
For all real numbers a and b,
$a + b = b + a$.
$ab = ba$.

Example: $5 + {}^-8 = {}^-8 + 5 = {}^-3$

Example: ${}^-2 \times 6 = 6 \times {}^-2 = {}^-12$

Associative
For all real numbers a, b, and c,
$(a + b) + c = a + (b + c)$.
$(ab)c = a(bc)$.

Example: $({}^-2 + 7) + 5 = {}^-2 + (7 + 5)$
$5 + 5 = {}^-2 + 12 = 10$

Additive Identity (Property of Zero)
There exists a unique real number 0 (zero) such that
$a + 0 = 0 + a = a$ for every real number a.

Example: $17 + 0 = 17$
The sum of any number and zero is that number.

Multiplicative Identity (Property of One)
There exists a unique nonzero real number 1 (one) such that
$1 \cdot a = a \cdot 1 = a$.

$a \cdot 1 = a$

Example: $^-34 \times 1 = {}^-34$
The product of any number and one is that number.

Additive Inverse (Property of Opposites)
For each real number a, there exists a real number $-a$ (the opposite of a) such that the sum of any number and its opposite is zero. $a + (-a) = (-a) + a = 0$.

Example: $25 + {}^-25 = 0$

Multiplicative Inverse (Property of Reciprocals)
For each nonzero real number, there exists a real number $1/a$ (the reciprocal of a) such that $a(1/a) = (1/a)a = 1$.

Example: $5 \times \frac{1}{5} = 1$

The product of any number and its reciprocal is one.

Distributive

$a (b + c) = ab + ac$

Example: $6 \times ({}^-4 + 9) = (6 \times {}^-4) + (6 \times 9)$
$6 \times 5 = {}^-24 + 54 = 30$

To multiply a sum by a number, multiply each addend by the number, then add the products.

Review of Law of Exponents: If a and b are real numbers and m and n are rational numbers, then,

1. $a^m \times a^n = a^{(m+n)}$

2. $\dfrac{a^m}{a^n} = a^{(m-n)}$

3. $(a^m)^n = a^{(mn)}$

4. $(ab)^m = a^m b^m$

5. $a^{-n} = \dfrac{1}{a^n} = (1/a)^n$

6. If a is any nonzero number, then $a^0 = 1$.

7. $a^{m/n} = \left(a^{1/n}\right)^m = \left(a^m\right)^{1/n}$

8. $\sqrt[n]{a^m}$ (radical form) $= a^{m/n}$ in exponential form.

Rules for operating with exponents:

1. To multiply two or more numbers that have the same base, we add the exponents and keep the same base.

$$2^3 \times 2^2 = 2^5 = 32$$

2. To divide two numbers that have the same base, we subtract the exponents and keep the same base.

$$\frac{3^4}{3^2} = 3^2 = 9$$

3. To change the sign of an exponent, the reciprocal of the number is used.

$$\left(\frac{3}{4}\right)^{-2} = \left(\frac{4}{3}\right)^2 = \frac{4}{3} \times \frac{4}{3} = \frac{16}{9}$$

4. The answer to any real number raised to a 0 exponent is 1.

$$3^0 = 1$$
$$(^-2)^0 = 1$$
$$^-3^0 = {}^- 1$$

5. Raising a number with an exponent to another exponent requires multiplying the exponents and keeping the same base.

$$(2^2)^3 = 2^6 = 64$$

Caution: Rules for exponents do not apply unless multiplying or dividing numbers with the same base.

Example 1:

$5^4 + 3^3$ Addition and bases are different, so multiply each number separately and combine the answers.

$$(5 \times 5 \times 5 \times 5) + (3 \times 3 \times 3) = 625 + 27 = 652$$

Distinguish relationships between the complex number system and its subsystems.

Complex numbers are of the form $a + b\,\mathbf{i}$, where a and b are real numbers and $\mathbf{i} = \sqrt{^-1}$. When $\mathbf{i}$ appears in an answer, it is acceptable unless it is in a denominator. When $\mathbf{i}^2$ appears in a problem, it is always replaced by a $^-1$. Remember, $\mathbf{i}^2 = ^-\mathbf{1}$.

To add or subtract complex numbers, add or subtract the real parts then add or subtract the imaginary parts and keep the i (just like combining like terms).

<u>Examples</u>: Add $(2 + 3i) + (^-7 - 4i)$.

$$2 + {}^-7 = {}^-5 \qquad 3i + {}^-4i = {}^-i \quad \text{so,}$$

$$(2 + 3i) + (^-7 - 4i) = {}^-5 - i$$

Subtract $(8 - 5i) - (^-3 + 7i)$

$$8 - 5i + 3 - 7i = 11 - 12i$$

To multiply 2 complex numbers, F.O.I.L. the 2 numbers together. Replace $\mathbf{i}^2$ with a $^-1$ and finish combining like terms. Answers should have the form $a + b\,\mathbf{i}$.

Example: Multiply $(8 + 3i)(6 - 2i)$ F.O.I.L. this.

When dividing 2 complex numbers, you must eliminate the complex number in the denominator.
If the complex number in the denominator is of the form $b\,i$, multiply both the numerator and denominator by **i**. Remember to replace i^2 with a $^-1$ and then continue simplifying the fraction.

Example:

$$\frac{2 + 3i}{5i} \qquad \text{Multiply this by } \frac{i}{i}$$

$$\frac{2 + 3i}{5i} \times \frac{i}{i} = \frac{(2 + 3i)\,i}{5i \cdot i} = \frac{2i + 3i^2}{5i^2} = \frac{2i + 3(^-1)}{^-5} = \frac{^-3 + 2i}{^-5} = \frac{3 - 2i}{5}$$

Complex numbers are numbers of the form $a + bi$, where a and b are real numbers and i is the imaginary unit $\sqrt{-1}$. Complex numbers attach meaning to and allow calculations with the square root of negative numbers. When graphing complex numbers, we plot the real number a on the x-axis (labeled R for real) and b on the y-axis (labeled I for imaginary). The modulus is the length of the **vector** from the origin to the position of the complex number on the graph. The argument is the angle the vector makes with the horizontal axis R.

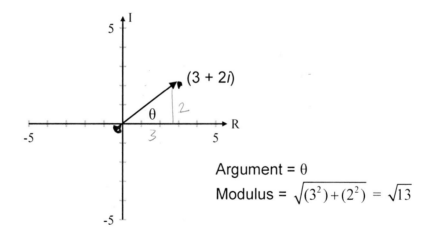

Argument = θ

Modulus = $\sqrt{(3^2) + (2^2)} = \sqrt{13}$

When graphing complex numbers in vector form, it is often desirable to present the numbers as **ordered pairs**. Thus, in the above example, the ordered pair of the plotted point is (3,2) when graphed in the complex plane.

An alternative representation of a complex number is the **polar form**. We can represent any complex number z with the following equation.

$z = r(\cos\theta + i\sin\theta)$,   where r is the modulus and θ is the argument.

Thus, from the example above, the polar form of $3 + 2i$ is

$$\sqrt{13}(\cos(\tan^{-1}\frac{2}{3}) + i\sin(\tan^{-1}\frac{2}{3})).$$

The final alternative representation of a complex number is the **exponential form**. We can represent any complex number z with the following equation:

$$z = re^{i\vartheta}$$

Thus, from the example above, the exponential form of $3 + 2i$ is

$$\sqrt{13}e^{0.588i}$$ (with the argument, θ, given in radians).

Writing complex numbers in polar form simplifies multiplication and division applications. Conversely, Cartesian notation ($a + bi$) makes addition and subtraction applications easier to manage. In addition, representing complex numbers as vectors enables addition and subtraction in graphical form. Finally, writing complex numbers in exponential form is often more convenient than polar form because it eliminates cumbersome trigonometric relations.

COMPETENCY 0003   UNDERSTAND ALGEBRAIC OPERATIONS AND PROPERTIES OF FUNCTIONS AND RELATIONS

Factor polynomials.

- To factor a polynomial, follow these steps:

a. **Factor out any GCF** (greatest common factor)

b. For a binomial (2 terms), check to see if the problem is the **difference of perfect squares**. If both factors are perfect squares, then it factors this way:

$$a^2 - b^2 = (a - b)(a + b)$$

If the problem is not the difference of perfect squares, then check to see if the problem is either the sum or difference of perfect cubes.

$$x^3 - 8y^3 = (x - 2y)(x^2 + 2xy + 4y^2) \qquad \leftarrow \text{difference}$$

$$64a^3 + 27b^3 = (4a + 3b)(16a^2 - 12ab + 9b^2) \quad \leftarrow \text{sum}$$

** The sum of perfect squares does NOT factor.

c. Trinomials could be perfect squares. Trinomials can be factored into 2 binomials (un-FOILing). Be sure the terms of the trinomial are in descending order. If last sign of the trinomial is a "+", then the signs in the parentheses will be the same as the sign in front of the second term of the trinomial. If the last sign of the trinomial is a "-", then there will be one "+" and one "-" in the two parentheses. The first term of the trinomial can be factored to equal the first terms of the two factors. The last term of the trinomial can be factored to equal the last terms of the two factors. Work backwards to determine the correct factors to multiply together to get the correct center term.

Factor completely:

1. $4x^2 - 25y^2$

2. $6b^2 - 2b - 8$

3. Find a factor of $6x^2 - 5x - 4$

    a. $(3x + 2)$ b. $(3x - 2)$ c. $(6x - 1)$ d. $(2x + 1)$

Answers:

1. No GCF; this is the difference of perfect squares.

$$4x^2 - 25y^2 = (2x - 5y)(2x + 5y)$$

2. GCF of 2; Try to factor into 2 binomials:

$$6b^2 - 2b - 8 = 2(3b^2 - b - 4)$$

Signs are one "$+$", one "$-$". $3b^2$ factors into $3b$ and b. Find factors of 4: 1 & 4; 2 & 2.

$$6b^2 - 2b - 8 = 2(3b^2 - b - 4) = 2(3b - 4)(b + 1)$$

3. If an answer choice is correct, find the other factor:

    a. $(3x + 2)(2x - 2) = 6x^2 - 2x - 4$

    b. $(3x - 2)(2x + 2) = 6x^2 + 2x - 4$

    c. $(6x - 1)(x + 4) = 6x^2 + 23x - 4$

    d. $(2x + 1)(3x - 4) = 6x^2 - 5x - 4$ $\leftarrow$ correct factors

Identify the slope and intercepts of a graph or an equation.

To find the y intercept, substitute 0 for x and solve for y. This is the y intercept. The y intercept is also the value of b in $y = mx + b$.

To find the x intercept, substitute 0 for y and solve for x. This is the x intercept.

1. Find the slope and intercepts of $3x + 2y = 14$.

$$3x + 2y = 14$$
$$2y = {}^{-}3x + 14$$
$$y = {}^{-}3/2\ x + 7$$

The slope of the line is $^{-}3/2$, the value of m.
The y intercept of the line is 7.

The intercepts can also be found by substituting 0 in place of the other variable in the equation.

To find the y intercept:
let $x = 0$; $3(0) + 2y = 14$
$0 + 2y = 14$
$2y = 14$
$y = 7$
$(0,7)$ is the y intercept.

To find the x intercept:
let $y = 0$; $3x + 2(0) = 14$
$3x + 0 = 14$
$3x = 14$
$x = 14/3$
$(14/3, 0)$ is the x intercept.

Find the slope and the intercepts (if they exist) for these equations:

1. $5x + 7y = {}^{-}70$
2. $x - 2y = 14$
3. $5x + 3y = 3(5 + y)$
4. $2x + 5y = 15$

Identify the interpretation of the slope and intercepts, given a real-world context.

Exercise: Interpreting Slope as a Rate of Change
Connection: Social Sciences/Geography

Real-life Application: Slope is often used to describe a constant or average rate of change. These problems usually involve units of measure such as miles per hour or dollars per year.

Problem:

The town of Verdant Slopes has been experiencing a boom in population growth. By the year 2000, the population had grown to 45,000, and by 2005, the population had reached 60,000.

Communicating about Algebra:

a. Using the formula for slope as a model, find the average rate of change in population growth, expressing your answer in people per year.

Extension:

b. Using the average rate of change determined in a., predict the population of Verdant Slopes in the year 2010.

Solution:

a. *Let t represent the time and p represent population growth. The two observances are represented by (t_1, p_1) and (t_2, p_2).*

1st observance = (t_1, p_1) = (2000, 45000)
2nd observance = (t_2, p_2) = (2005, 60000)

Use the formula for slope to find the average rate of change.

$$\text{Rate of change} = \frac{p_2 - p_1}{t_2 - t_1}$$

Substitute values.

$$= \frac{60000 - 45000}{2005 - 2000}$$

Simplify.

$$= \frac{15000}{5} = 3000 \, people \, / \, year$$

The average rate of change in population growth for Verdant Slopes between the years 2000 and 2005 was 3000 people/year.

b.

$$3000 \, people \, / \, year \times 5 \, years = 15000 \, people$$
$$60000 \, people + 15000 \, people = 75000 \, people$$

At a continuing average rate of growth of 3000 people/year, the population of Verdant Slopes could be expected to reach 75,000 by the year 2010.

Solve equations involving radicals, limited to square roots.

To solve a radical equation:

1. Get a term with a radical alone on one side of the equation.

2. Raise both sides of the equation to a power equal to the index on the radical (that is, square both sides of an equation containing a square root. Cube both sides of an equation containing a cube root etc.). DO NOT SQUARE (OR CUBE, etc.) each term separately. SQUARE (OR CUBE, etc.) the ENTIRE SIDE of the equation.

3. If there are any radicals remaining, repeat steps 1 and 2 until all radicals are gone.

4. Solve the remaining equation.

5. Check your answers in the original radical equation. Every answer may not check. If no answer checks in the original equation, then the answer to the equation is $\varnothing$, the empty set or null set.

NOTE: Since answers on this type of problem must be checked in the original equation anyhow, this means that possible answers on a multiple choice test could be immediately substituted into the problem to find the correct answer without having to solve the actual problem.

Solve and check:

$\sqrt{2x-8} - 7 = 9$ Get radical alone.

$\sqrt{2x-8} = 16$ Square both sides.

$\sqrt{2x-8}^2 = 16^2$

$2x - 8 = 256$ Solve for x.

$2x = 264$

$x = 132$

Check:

$\sqrt{2(132)-8} - 7 = 9$

$\sqrt{264-8} - 7 = 9$

$\sqrt{256} - 7 = 9$

$16 - 7 = 9$ This answer checks.

Solve and check:

$\sqrt{5x-1} - 1 = x$ Add 1.

$\left(\sqrt{5x-1}\right)^2 = (x+1)^2$ Square both sides.

$5x - 1 = x^2 + 2x + 1$ Solve this equation.

$0 = x^2 - 3x + 2$

$0 = (x-2)(x-1)$

$x = 2$ $x = 1$

Check both answers:

$\sqrt{5(2)-1} - 1 = 2$ $\sqrt{5(1)-1} - 1 = 1$

$3 - 1 = 2$ $2 - 1 = 1$

When these are checked, both answers check.

Identify the domain and range of specified functions.

- The **domain** of a relation is the set made of all the first coordinates of the ordered pairs.

- The **range** of a relation is the set made of all the second coordinates of the ordered pairs.

- A **function** is a relation in which different ordered pairs have different first coordinates. (No x values are repeated.)

- A **mapping** is a diagram with arrows drawn from each element of the domain to the corresponding elements of the range. If 2 arrows are drawn from the same element of the domain, then it is not a function.

Example:

Determine the domain and range of this mapping.

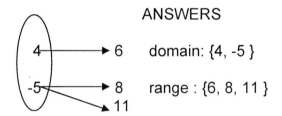

ANSWERS

domain: {4, -5 }

range : {6, 8, 11 }

Practice Problems:

1.. Determine the domain and range of this graph.

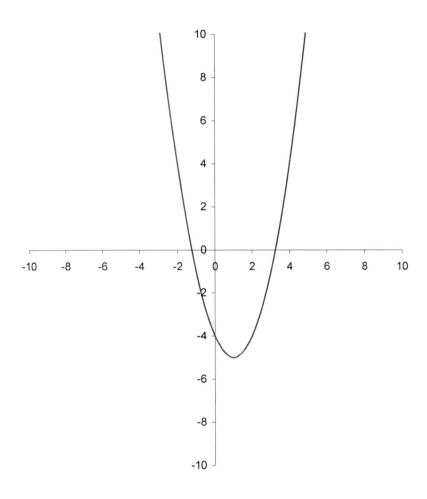

2. If $A = \{(x, y) \mid y = x^2 - 6\}$, find the domain and range.

3. Give the domain and range of set B if:

$$B = \{(1, {}^-2), (4, {}^-2), (7, {}^-2), (6, {}^-2)\}$$

4. Determine the domain of this function:

$$f(x) = \frac{5x + 7}{x^2 - 4}$$

5. Determine the domain and range of these graphs.

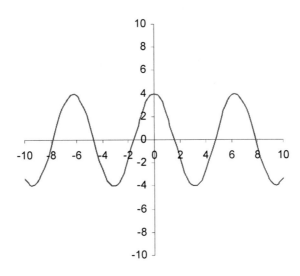

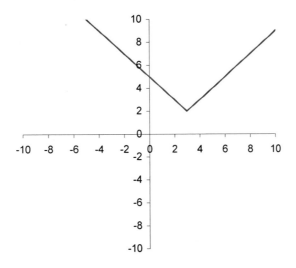

6. If $E = \{(x, y) \mid y = 5\}$, find the domain and range.

7. Determine the ordered pairs in the relation shown in this mapping.

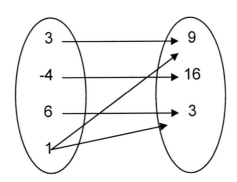

Different types of function transformations affect the graph and characteristics of a function in predictable ways. The basic types of transformation are horizontal and vertical shift (translation), horizontal and vertical scaling (dilation), and reflection. As an example of the types of transformations, we will consider transformations of the functions $f(x) = x^2$.

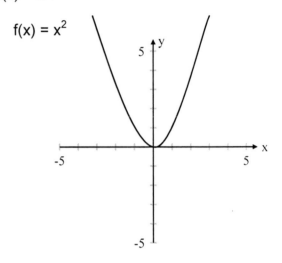

$f(x) = x^2$

Horizontal shifts take the form $g(x) = f(x \pm c)$. For example, we obtain the graph of the function $g(x) = (x + 2)^2$ by shifting the graph of $f(x) = x^2$ two units to the left. The graph of the function $h(x) = (x - 2)^2$ is the graph of $f(x) = x^2$ shifted two units to the right.

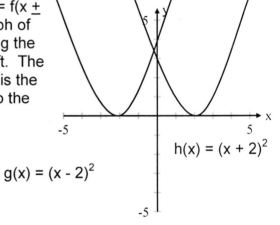

$h(x) = (x + 2)^2$

$g(x) = (x - 2)^2$

Vertical shifts take the form $g(x) = f(x) \pm c$. example, we obtain the graph of the function $g(x) = (x^2) - 2$ by shifting the graph of $f(x) = x^2$ two units down. The graph of the function $h(x) = (x^2) + 2$ is the graph of $f(x) = x^2$ shifted two units up.

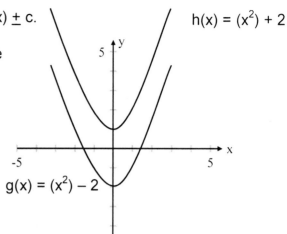

Horizontal scaling takes the form $g(x) = f(cx)$. For example, we obtain the graph of the function $g(x) = (2x)^2$ by compressing the graph of $f(x) = x^2$ in the x-direction by a factor of two. If $c > 1$ the graph is compressed in the x-direction, while if $1 > c > 0$ the graph is stretched in the x-direction.

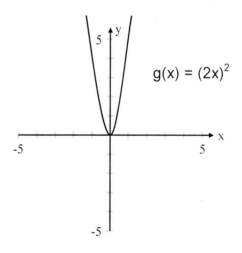

Determine the inverse of a given function.

How to write the equation of the inverse of a function.

1. To find the inverse of an equation using x and y, replace each letter with the other letter. Then solve the new equation for y, when possible. Given an equation like $y = 3x - 4$, replace each letter with the other:

$$x = 3y - 4.$$ Now solve this for y:
$$x + 4 = 3y$$
$$1/3\, x + 4/3 = y \quad \text{This is the inverse.}$$

Sometimes the function is named by a letter:

$$f(x) = 5x + 10$$

Temporarily replace f(x) with y.

$$y = 5x + 10$$

Now replace each letter with the other: $x = 5y + 10$
Solve for the new y: $x - 10 = 5y$
$$1/5\, x - 2 = y$$

The inverse of f(x) is denoted as f$^{-1}(x)$, so the answer is

$$f^{-1}(x) = 1/5\, X - 2\ .$$

Determine the composition of two functions.

Composition is a process that creates a new function by substituting an entire function into another function. The composition of two functions f(x) and g(x) is denoted by (f ∘ g)(x) or f(g(x)). The domain of the composed function, f(g(x)), is the set of all values of x in the domain of g that produce a value for g(x) that is in the domain of f. In other words, f(g(x)) is defined whenever both g(x) and f(g(x)) are defined.

Example 1:

If f(x) = x + 1 and g(x) = x^3, find the composition functions f ∘ g and g ∘ f and state their domains.

Solution:

(f ∘ g)(x) = f(g(x)) = f(x^3) = x^3 + 1
(g ∘ f)(x) = g(f(x)) = g(x + 1) = (x + 1)^3

The domain of both composite functions is the set of all real numbers.

Note that f(g(x)) and g(f(x)) are not the same. In general, unlike multiplication and addition, composition is not reversible. Thus, the order of composition is important.

Example 2:

If f(x) = sqrt(x) and g(x) = x +2, find the composition functions f ∘ g and g ∘ f and state their domains.

Solution:

(f ∘ g)(x) = f(g(x)) = f(x + 2) = sqrt(x + 2)
(g ∘ f)(x) = g(f(x)) = g(sqrt(x)) = sqrt(x) +2

The domain of f(g(x)) is x ≥ -2 because x + 2 must be non-negative in order to take the square root.

The domain of g(f(x)) is x ≥ 0 because x must be non-negative in order to take the square root.

Note that defining the domain of composite functions is important when square roots are involved.

Find a specified term in an arithmetic sequence.

When given a set of numbers where the common difference between the terms is constant, use the following formula:

$$a_n = a_1 + (n-1)d \quad \text{where } a_1 = \text{the first term}$$

$$n = \text{the } n \text{ th term (general term)}$$
$$d = \text{the common difference}$$

Sample problem:

1. Find the 8th term of the arithmetic sequence 5, 8, 11, 14, ...

$a_n = a_1 + (n-1)d$	
$a_1 = 5$	Identify 1st term.
$d = 3$	Find d.
$a_8 = 5 + (8-1)3$	Substitute.
$a_8 = 26$	

2. Given two terms of an arithmetic sequence find a and d.

$$a_4 = 21 \qquad a_6 = 32$$
$$a_n = a_1 + (n-1)d$$
$$21 = a_1 + (4-1)d$$
$$32 = a_1 + (6-1)d$$

$21 = a_1 + 3d$	Solve the system of equations.
$32 = a_1 + 5d$	

$21 = a_1 + 3d$	
$-32 = \ ^-a_1 - 5d$	Multiply by $^-1$ and add the equations.
$^-11 = \ ^-2d$	
$5.5 = d$	

$21 = a_1 + 3(5.5)$	Substitute $d = 5.5$ into one of the equations.
$21 = a_1 + 16.5$	
$a_1 = 4.5$	

The sequence begins with 4.5 and has a common difference of 5.5 between numbers.

Find a specified term in a geometric sequence.

When using geometric sequences consecutive numbers are compared to find the common ratio.

$$r = \frac{a_{n+1}}{a_n}$$

r = the common ratio
a_n = the nth term

The ratio is then used in the geometric sequence formula:

$$a_n = a_1 r^{n-1}$$

Sample problems:

1. Find the 8th term of the geometric sequence 2, 8, 32, 128 ...

$r = \dfrac{a_{n+1}}{a_n}$ Use the common ratio formula to find r.

$r = \dfrac{8}{2} = 4$ Substitute $a_n = 2$ $a_{n+1} = 8$

$a_n = a_1 \times r^{n-1}$ Use $r = 4$ to solve for the 8th term.
$a_8 = 2 \times 4^{8-1}$
$a_8 = 32768$

COMPETENCY 0004   UNDERSTAND PROPERTIES OF LINEAR EQUATIONS AND LINEAR SYSTEMS.

Analyze relationships between tables, graphs, or rules.

A relationship between two quantities can be shown using a table, graph or rule. In this example, the rule y= 9x describes the relationship between the total amount earned, y, and the total amount of $9 sunglasses sold, x.

A table using this data would appear as:

number of sunglasses sold	1	5	10	15
total dollars earned	9	45	90	135

Each *(x,y)* relationship between a pair of values is called the coordinate pair and can be plotted on a graph. The coordinate pairs *(1,9)*, *(5,45)*, *(10,90)*, and *(15,135)*, are plotted on the graph below.

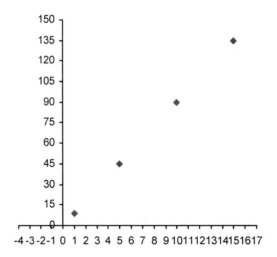

The graph above shows a linear relationship. A linear relationship is one in which two quantities are proportional to each other. Doubling *x* also doubles *y*. On a graph, a straight line depicts a linear relationship.

Another type of relationship is a nonlinear relationship. This is one in which change in one quantity does not affect the other quantity to the same extent. Nonlinear graphs have a curved line such as the graph below.

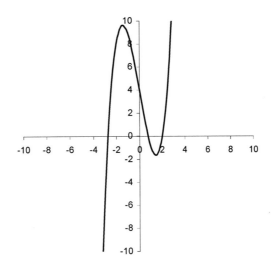

Determine an equation of a line.

Slope – The slope of a line is the "slant" of a line. A downward left to right slant means a negative slope. An upward slant is a positive slope. The formula for calculating the slope of a line with coordinates $(x_1, y_1) and (x_2, y_2)$ is:

$$\text{slope} = \frac{y_2 - y_1}{x_2 - x_1}$$

The top of the fraction represents the change in the y coordinates; it is called the **rise**.

The bottom of the fraction represents the change in the x coordinates, it is called the **run.**

Example: Find the slope of a line with points at (2,2) and (7,8).

$\frac{(8) - (2)}{(7) - (2)}$ plug the values into the formula

$\frac{6}{5}$ solve the rise over run

= 1.2   solve for the slope

The y intercept is the y coordinate of the point where a line crosses the y axis. The equation can be written in slope-intercept form, which is $y = mx + b$, where m is the slope and b is the y intercept.

To rewrite the equation into some other form, multiply each term by the common denominator of all the fractions. Then rearrange terms as necessary.

The equation of a line from its graph can be found by finding its slope (see Skill 3.2 for the slope formula) and its *y* intercept.

$$Y - y_a = m(X - x_a)$$

(x_a, y_a) can be (x_1, y_1) or (x_2, y_2) If **m**, the value of the slope, is distributed through the parentheses, the equation can be rewritten into other forms of the equation of a line.

Example: Find the equation of a line through $(9, {}^-6)$ and $({}^-1, 2)$.

$$\text{slope} = \frac{y_2 - y_1}{x_2 - x_1} = \frac{2 - {}^-6}{{}^-1 - 9} = \frac{8}{{}^-10} = -\frac{4}{5}$$

$$Y - y_a = m(X - x_a) \rightarrow Y - 2 = {}^-4/5(X - {}^-1) \rightarrow$$

$$Y - 2 = {}^-4/5(X + 1) \rightarrow Y - 2 = {}^-4/5\,X - 4/5 \rightarrow$$

$$Y = {}^-4/5\;X + 6/5 \quad \text{This is the slope-intercept form.}$$

Multiplying by 5 to eliminate fractions, it is:

$$5Y = {}^-4X + 6 \rightarrow 4X + 5Y = 6 \quad \text{Standard form.}$$

Identify matrices that represent data provided by real-world or mathematical problems.

A **matrix** is an ordered set of numbers in rectangular form.

$$\begin{pmatrix} 0 & 3 & 1 \\ 4 & 2 & 3 \\ 1 & 0 & 2 \end{pmatrix}$$

Since this matrix has 3 rows and 3 columns, it is called a 3 x 3 matrix. The element in the second row, third column would be denoted as $3_{2,3}$.

Matrices are used often to solve systems of equations. They are also used by physicists, mathematicians, and biologists to organize and study data such as population growth. It is also used in finance for such purposes as investment growth analysis and portfolio analysis. Matrices are easily translated into computer code in high-level programming languages and can be easily expressed in electronic spreadsheets.

A simple financial example of using a matrix to solve a problem follows:

A company has two stores. The income and expenses (in dollars) for the two stores, for three months, are shown in the matrices.

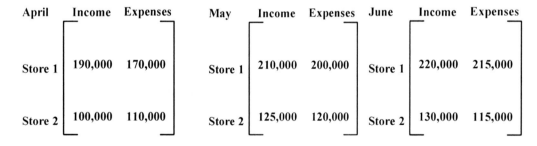

April	Income	Expenses		May	Income	Expenses		June	Income	Expenses
Store 1	190,000	170,000		Store 1	210,000	200,000		Store 1	220,000	215,000
Store 2	100,000	110,000		Store 2	125,000	120,000		Store 2	130,000	115,000

The owner wants to know what his first-quarter income and expenses were, so he adds the three matrices.

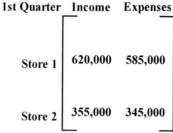

Then, to find the profit for each store:

Profit for Store 1 = $620,000 - $585,000 = $35,000
Profit for Store 2 = $355,000 - $345,000 = $10,000

Identify graphs of linear equations or inequalities involving **two variables on the coordinate plane.**

A first degree equation has an equation of the form $ax + by = c$. To graph this equation, find either one point and the slope of the line or find two points. To find a point and slope, solve the equation for y. This gets the equation in **slope intercept form**, $y = mx + b$. The point (0,b) is the y-intercept and m is the line's slope. To find any 2 points, substitute any 2 numbers for x and solve for y. To find the intercepts, substitute 0 for x and then 0 for y.

Remember that graphs will go up as they go to the right when the slope is positive. Negative slopes make the lines go down as they go to the right.

If the equation solves to x = **any number**, then the graph is a **vertical line**. It only has an x intercept. Its slope is **undefined**.

If the equation solves to y = **any number**, then the graph is a **horizontal line**. It only has a y intercept. Its slope is 0 (zero).

- When graphing a linear inequality, the line will be dotted if the inequality sign is < or >. If the inequality signs are either ≥ or ≤, the line on the graph will be a solid line. Shade above the line when the inequality sign is ≥ or >. Shade below the line when the inequality sign is < or ≤. Inequalities of the form $x >, x \leq, x <,$ or $x \geq$ number, draw a vertical line (solid or dotted). Shade to the right for > or ≥. Shade to the left for < or ≤. Remember: **Dividing or multiplying by a negative number will reverse the direction of the inequality sign.**

Identify equations or inequalities that could be used to solve real-world and mathematical problems involving one or two variables.

Word problems can sometimes be solved by using a system of two equations in 2 unknowns. This system can then be solved using substitution, the addition-subtraction method, or graphing.

Example: Mrs. Winters bought 4 dresses and 6 pairs of shoes for $340. Mrs. Summers went to the same store and bought 3 dresses and 8 pairs of shoes for $360. If all the dresses were the same price and all the shoes were the same price, find the price charged for a dress and for a pair of shoes.

Let x = price of a dress
Let y = price of a pair of shoes

Then Mrs. Winters' equation would be: $4x + 6y = 340$
Mrs. Summers' equation would be: $3x + 8y = 360$

To solve by addition-subtraction:

Multiply the first equation by 4: $4(4x + 6y = 340)$

Multiply the other equation by $^-3$: $^-3(3x + 8y = 360)$
By doing this, the equations can be added to each other to eliminate one variable and solve for the other variable.

$$16x + 24y = 1360$$
$$\underline{-9x - 24y = {}^-1080}$$
$$7x = 280$$
$$x = 40 \leftarrow \text{the price of a dress was \$40}$$

solving for y, $y = 30 \leftarrow$ the price of a pair of shoes,$30

Example: Aardvark Taxi charges $4 initially plus $1 for every mile traveled. Baboon Taxi charges $6 initially plus $.75 for every mile traveled. Determine when it is cheaper to ride with Aardvark Taxi or to ride with Baboon Taxi.

Aardvark Taxi's equation: $y = 1x + 4$

Baboon Taxi's equation : $y = .75x + 6$

Using substitution: $.75x + 6 = x + 4$

Multiplying by 4: $3x + 24 = 4x + 16$

Solving for x : $8 = x$

This tells you that at 8 miles the total charge for the 2 companies is the same. If you compare the charge for 1 mile, Aardvark charges $5 and Baboon charges $6.75. Clearly Aardvark is cheaper for distances up to 8 miles, but Baboon Taxi is cheaper for distances greater than 8 miles.

This problem can also be solved by graphing the 2 equations.

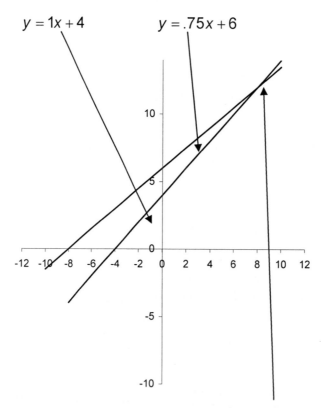

$y = 1x + 4$ $y = .75x + 6$

The lines intersect at (8, 12), therefore at 8 miles both companies charge $12. At values less than 8 miles, Aardvark Taxi charges less (the graph is below Baboon). Greater than 8 miles, Aardvark charges more (the graph is above Baboon).

Equations and inequalities can be used to solve various types of word problems. Examples follow.

Example: The YMCA wants to sell raffle tickets to raise at least $32,000. If they must pay $7,250 in expenses and prizes out of the money collected from the tickets, how many tickets worth $25 each must they sell?

Solution: Since they want to raise **at least $32,000**, that means they would be happy to get 32,000 **or more**. This requires an inequality.

Let x = number of tickets sold
Then $25x$ = total money collected for x tickets

Total money minus expenses is greater than $32,000.

$$25x - 7250 \geq 32000$$
$$25x \geq 39250$$
$$x \geq 1570$$

If they sell **1,570 tickets or more**, they will raise AT LEAST $32,000.

Solve systems of linear equations involving two variables using graphing, substitution, or elimination.

Word problems can sometimes be solved by using a system of two equations in 2 unknowns. This system can then be solved using **substitution**, or the **addition-subtraction method**.

Example: Farmer Greenjeans bought 4 cows and 6 sheep for $1700. Mr. Ziffel bought 3 cows and 12 sheep for $2400. If all the cows were the same price and all the sheep were another price, find the price charged for a cow or for a sheep.

Let x = price of a cow
Let y = price of a sheep

Then Farmer Greenjeans' equation would be: $4x + 6y = 1700$
Mr. Ziffel's equation would be: $3x + 12y = 2400$

To solve by **addition-subtraction**:

Multiply the first equation by $^-2$: $^-2(4x + 6y = 1700)$
Keep the other equation the same : $(3x + 12y = 2400)$
By doing this, the equations can be added to each other to eliminate one variable and solve for the other variable.

$$^-8x - 12y = ^-3400$$
$$\underline{3x + 12y = 2400} \qquad \text{Add these equations.}$$
$$^-5x \qquad = ^-1000$$

$x = 200 \leftarrow$ the price of a cow was $200.
Solving for y, $y = 150 \leftarrow$ the price of a sheep, $150.

To solve by **substitution**:

Solve one of the equations for a variable. (Try to make an equation without fractions if possible.) Substitute this expression into the equation that you have not yet used. Solve the resulting equation for the value of the remaining variable.

$$4x + 6y = 1700$$
$$3x + 12y = 2400 \leftarrow \text{Solve this equation for } x.$$

It becomes $x = 800 - 4y$. Now substitute $800 - 4y$ in place of x in the OTHER equation. $4x + 6y = 1700$ now becomes:

$$4(800 - 4y) + 6y = 1700$$
$$3200 - 16y + 6y = 1700$$
$$3200 - 10y = 1700$$
$$^-10y = ^-1500$$
$$y = 150, \text{ or } \$150 \text{ for a sheep.}$$

Substituting 150 back into an equation for y, find x.

$$4x + 6(150) = 1700$$
$$4x + 900 = 1700$$
$$4x = 800 \text{ so } x = 200 \text{ for a cow.}$$

Word problems can sometimes be solved by using a system of three equations in 3 unknowns. This system can then be solved using **substitution** or the **addition-subtraction method**.

To solve by **substitution**:

Example: Mrs. Allison bought 1 pound of potato chips, a 2 pound beef roast, and 3 pounds of apples for a total of $ 8.19. Mr. Bromberg bought a 3 pound beef roast and 2 pounds of apples for $ 9.05. Kathleen Kaufman bought 2 pounds of potato chips, a 3 pound beef roast, and 5 pounds of apples for $ 13.25. Find the per pound price of each item.

Let x = price of a pound of potato chips
Let y = price of a pound of roast beef
Let z = price of a pound of apples

Mrs. Allison's equation would be: $1x + 2y + 3z = 8.19$
Mr. Bromberg's equation would be: $3y + 2z = 9.05$
K. Kaufman's equation would be: $2x + 3y + 5z = 13.25$

Take the first equation and solve it for x. (This was chosen because x is the easiest variable to get alone in this set of equations.) This equation would become:

$$x = 8.19 - 2y - 3z$$

Substitute this expression into the other equations in place of the letter x:

$$3y + 2z = 9.05 \leftarrow \text{ equation 2}$$
$$2(8.19 - 2y - 3z) + 3y + 5z = 13.25 \leftarrow \text{ equation 3}$$

Simplify the equation by combining like terms:

$$3y + 2z = 9.05 \leftarrow \text{ equation 2}$$
$$* \ ^{-}1y - 1z = \ ^{-}3.13 \leftarrow \text{equation 3}$$

Solve equation 3 for either y or z:

$y = 3.13 - z$ Substitute this into equation 2 for y:

$$3(3.13 - z) + 2z = 9.05 \leftarrow \text{ equation 2}$$
$$^{-}1y - 1z = \ ^{-}3.13 \leftarrow \text{ equation 3}$$

Combine like terms in equation 2:

$$9.39 - 3z + 2z = 9.05$$
$$z = .34 \quad \text{per pound price of apples}$$

Substitute .34 for z in the starred equation above to solve for y:
$$y = 3.13 - z \quad \text{becomes} \quad y = 3.13 - .34, \text{ so}$$
$$y = 2.79 \quad = \text{per pound price of roast beef}$$

Substituting .34 for z and 2.79 for y in one of the original equations, solve for x:

$$1x + 2y + 3z = 8.19$$
$$1x + 2(2.79) + 3(.34) = 8.19$$
$$x + 5.58 + 1.02 = 8.19$$
$$x + 6.60 = 8.19$$
$$x = 1.59 \quad \text{per pound of potato chips}$$

$$(x, y, z) = (\ 1.59,\ 2.79,\ .34)$$

To solve by **addition-subtraction**:

Choose a letter to eliminate. Since the second equation is already missing an x, let's eliminate x from equations 1 and 3.

1) $1x + 2y + 3x = 8.19 \leftarrow$ Multiply by $^-2$ below.
2) $\qquad 3y + 2z = 9.05$
3) $2x + 3y + 5z = 13.25$

$^-2(1x + 2y + 3z = 8.19) \quad = \quad ^-2x - 4y - 6z = \ ^-16.38$
Keep equation 3 the same : $\quad 2x + 3y + 5z = 13.25$

By doing this, the equations $\qquad\qquad ^-y - z = \ ^-3.13 \leftarrow$ equation 4
can be added to each other to
eliminate one variable.

The equations left to solve are equations 2 and 4:
$$^-y - z = \ ^-3.13 \leftarrow \text{equation 4}$$
$$3y + 2z = 9.05 \leftarrow \text{equation 2}$$

Multiply equation 4 by 3: $3(^-y - z = ^-3.13)$
Keep equation 2 the same: $3y + 2z = 9.05$

$$^-3y - 3z = ^-9.39$$
$$\underline{3y + 2z = 9.05} \qquad \text{Add these equations.}$$
$$^-1z = ^-.34$$

$z = .34 \leftarrow$ the per pound price of apples
solving for y, $y = 2.79 \leftarrow$ the per pound roast beef price
solving for x, $x = 1.59 \leftarrow$ potato chips, per pound price

To solve by **substitution**:

Solve one of the 3 equations for a variable. (Try to make an equation without fractions if possible.) Substitute this expression into the other 2 equations that you have not yet used.

1) $1x + 2y + 3z = 8.19 \leftarrow$ Solve for x.
2) $3y + 2z = 9.05$
3) $2x + 3y + 5z = 13.25$
Equation 1 becomes $x = 8.19 - 2y - 3z$.

Substituting this into equations 2 and 3, they become:

2) $3y + 2z = 9.05$
3) $2(8.19 - 2y - 3z) + 3y + 5z = 13.25$
    $16.38 - 4y - 6z + 3y + 5z = 13.25$
    $^-y - z = ^-3.13$
The equations left to solve are :

$3y + 2z = 9.05$

$^-y - z = ^-3.13 \leftarrow$ Solve for either y or z.

It becomes $y = 3.13 - z$. Now substitute $3.13 - z$ in place of y in the OTHER equation. $3y + 2z = 9.05$ now becomes:
$$3(3.13 - z) + 2z = 9.05$$
$$9.39 - 3z + 2z = 9.05$$
$$9.39 - z = 9.05$$
$$^-z = ^-.34$$
$$z = .34 \text{ , or } \$.34/\text{lb of apples}$$

Substituting .34 back into an equation for z, find y.

$$3y + 2z = 9.05$$
$$3y + 2(.34) = 9.05$$
$$3y + .68 = 9.05 \text{ so } y = 2.79/\text{lb of roast beef}$$

Substituting .34 for z and 2.79 for y into one of the original equations, it becomes:

$$2x + 3y + 5z = 13.25$$
$$2x + 3(2.79) + 5(.34) = 13.25$$
$$2x + 8.37 + 1.70 = 13.25$$
$$2x + 10.07 = 13.25, \text{ so } x = 1.59/\text{lb of potato chips}$$

Determine the solution set of a system of linear inequalities involving two variables.

To graph an inequality, solve the inequality for y. This gets the inequality in **slope intercept form**, (for example : $y < mx + b$). The point (0,b) is the y-intercept and m is the line's slope.

If the inequality solves to $x \geq$ **any number**, then the graph includes a **vertical line**.

If the inequality solves to $y \leq$ **any number**, then the graph includes a **horizontal line**.

When graphing a linear inequality, the line will be dotted if the inequality sign is $<$ or $>$. If the inequality signs are either $\geq$ or $\leq$, the line on the graph will be a solid line. Shade above the line when the inequality sign is $\geq$ or $>$. Shade below the line when the inequality sign is $<$ or $\leq$. For inequalities of the forms $x >$ number, $x \leq$ number, $x <$ number, or $x \geq$ number, draw a vertical line (solid or dotted). Shade to the right for $>$ or $\geq$. Shade to the left for $<$ or $\leq$.

Use these rules to graph and shade each inequality. The solution to a system of linear inequalities consists of the part of the graph that is shaded for each inequality. For instance, if the graph of one inequality was shaded with red, and the graph of another inequality was shaded with blue, then the overlapping area would be shaded purple. The purple area would be the points in the solution set of this system.

Example: Solve by graphing:

$$x + y \leq 6$$
$$x - 2y \leq 6$$

Solving the inequalities for y, they become:

$y \leq {}^-x + 6$ (y intercept of 6 and slope = $^-1$)

$y \geq 1/2x - 3$ (y intercept of $^-3$ and slope = $1/2$)

A graph with shading is shown below:

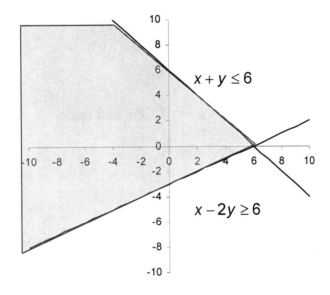

Resolve a vector into component vectors.

Occasionally, it is important to reverse the addition or subtraction process and express the single vector as the sum or difference of two other vectors. It may be critically important for a pilot to understand not only the air velocity but also the ground speed and the climbing speed.

Sample problem:

A pilot is traveling at an air speed of 300 mph and a direction of 20 degrees. Find the horizontal vector (ground speed) and the vertical vector (climbing speed).

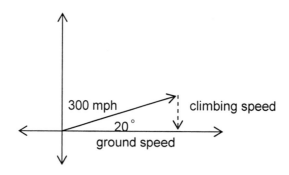

1. Draw sketch.

2. Use appropriate trigonometric ratio to calculate the component vectors.

To find the vertical vector:	To find the horizontal vector:
$\sin x = \dfrac{\text{opposite}}{\text{hypotenuse}}$	$\cos x = \dfrac{\text{adjacent}}{\text{hypotenuse}}$
$\sin(20) = \dfrac{c}{300}$	$\cos(20) = \dfrac{g}{300}$
$c = (.3420)(300)$	$g = (.9397)(300)$
$c = 102.606$	$g = 281.908$

Add or subtract vectors.

Vectors are used to measure displacement of an object or force.

Addition of vectors:

$$(a,b)+(c,d)=(a+c,b+d)$$

Addition Properties of vectors:

$$a+b=b+a$$
$$a+(b+c)=(a+b)+c$$
$$a+0=a$$
$$a+(^-a)=0$$

Subtraction of vectors:

$$a-b=a+(^-b) \text{ therefore,}$$
$$a-b=(a_1,a_2)+(^-b_1,^-b_2) \text{ or}$$
$$a-b=(a_1-b_1,a_2-b_2)$$

Sample problem:

If $a=(4,^-1)$ and $b=(^-3,6)$, find $a+b$ and $a-b$.

Using the rule for addition of vectors:

$$(4,^-1)+(^-3,6)=(4+(^-3),^-1+6)$$
$$=(1,5)$$

Using the rule for subtraction of vectors:

$$(4,^-1)-(^-3,6)=(4-(^-3),^-1-6)$$
$$=(7,^-7)$$

Find the dot product of two vectors.

The dot product $a \cdot b$:

$$a = (a_1, a_2) = a_1 i + a_2 j \quad \text{and} \quad b = (b_1, b_2) = b_1 i + b_2 j$$

$$a \cdot b = a_1 b_1 + a_2 b_2$$

$a \cdot b$ is read "a dot b". Dot products are also called scalar or inner products. When discussing dot products, it is important to remember that "a dot b" is not a vector, but a real number.

Properties of the dot product:

$$a \cdot a = |a|^2$$
$$a \cdot b = b \cdot a$$
$$a \cdot (b + c) = a \cdot b + a \cdot c$$
$$(ca) \cdot b = c(a \cdot b) = a \cdot (cb)$$
$$0 \cdot a = 0$$

Solve problems involving vectors.

Vectors are used often in navigation of ships and aircraft as well as in force and work problems.

Sample problem:

1. An airplane is flying with a heading of 60 degrees east of north at 450 mph. A wind is blowing at 37 mph from the north. Find the plane's ground speed to the nearest mph and direction to the nearest degree.

Draw a sketch.

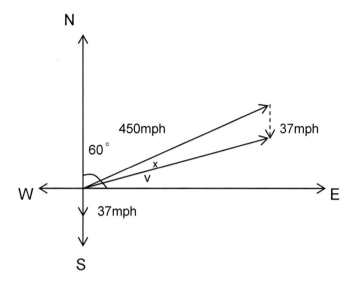

Use the law of cosines to find the ground speed.

$$a^2 = b^2 + c^2 - 2bc \cos A$$

$$|v|^2 = 450^2 + 37^2 - 2(450)(37) \cos 60$$

$$|v|^2 = 187,219$$

$$|v| = 432.688$$

$$|v| \approx 433 \text{ mph}$$

Use the law of sines to find the measure of x.

$$\frac{\sin A}{a} = \frac{\sin C}{c}$$

$$\frac{\sin 60}{433} = \frac{\sin x}{37}$$

$$\sin x = \frac{37(\sin 60)}{433}$$

$$x = 4.24$$

$$x \approx 4 \text{ degrees}$$

The plane's actual course is $60 + 4 = 64$ degrees east of north and the ground speed is approximately 433 miles per hour.

COMPETENCY 0005   UNDERSTAND PROPERTIES OF QUADRATIC FUNCTIONS.

Solve quadratic equations and inequalities by completing the square, the quadratic formula, and factoring.

A **quadratic equation** is written in the form $ax^2 + bx + c = 0$. To solve a quadratic equation by factoring, at least one of the factors must equal zero.

Example:
Solve the equation.

$x^2 + 10x - 24 = 0$

$(x + 12)(x - 2) = 0$ Factor.

$x + 12 = 0$ or $x - 2 = 0$ Set each factor equal to 0.

$x = {}^-12$ $x = 2$ Solve.

Check:

$x^2 + 10x - 24 = 0$

$({}^-12)^2 + 10({}^-12) - 24 = 0$ $(2)^2 + 10(2) - 24 = 0$

$144 - 120 - 24 = 0$ $4 + 20 - 24 = 0$

$0 = 0$ $0 = 0$

A quadratic equation that cannot be solved by factoring can be solved by **completing the square**.

Example:

Solve the equation.

$x^2 - 6x + 8 = 0$

$x^2 - 6x = {}^-8$ Move the constant to the right side.

$x^2 - 6x + 9 = {}^-8 + 9$ Add the square of half the coefficient of x to both sides.

$(x - 3)^2 = 1$ Write the left side as a perfect square.

$x - 3 = \pm\sqrt{1}$ Take the square root of both sides.

$x - 3 = 1$ $x - 3 = {}^-1$ Solve.

$x = 4$ $x = 2$

Check:

$$x^2 - 6x + 8 = 0$$

$$4^2 - 6(4) + 8 = 0 \qquad\qquad 2^2 - 6(2) + 8 = 0$$

$$16 - 24 + 8 = 0 \qquad\qquad 4 - 12 + 8 = 0$$

$$0 = 0 \qquad\qquad\qquad\quad 0 = 0$$

The general technique for graphing quadratics is the same as for graphing linear equations. Graphing quadratic equations, however, results in a parabola instead of a straight line.

Example:

Graph $y = 3x^2 + x - 2$.

x	$y = 3x^2 + x - 2$
$^-2$	8
$^-1$	0
0	$^-2$
1	2
2	12

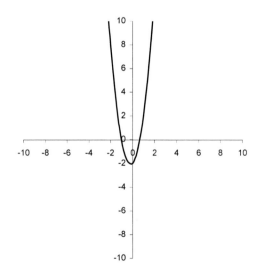

To solve a quadratic equation using the quadratic formula, be sure that your equation is in the form $ax^2 + bx + c = 0$. Substitute these values into the formula:

$$y = -\frac{1}{23}x + \frac{2}{23} - 15 = -\frac{1}{23}x - 14\frac{21}{23}$$

<u>Example</u>:

Solve the equation.

$$3x^2 = 7 + 2x \rightarrow 3x^2 - 2x - 7 = 0$$

$$a = 3 \quad b = {}^-2 \quad c = {}^-7$$

$$x = \frac{-({}^-2) \pm \sqrt{({}^-2)^2 - 4(3)({}^-7)}}{2(3)}$$

$$x = \frac{2 \pm \sqrt{4 + 84}}{6}$$

$$x = \frac{2 \pm \sqrt{88}}{6}$$

$$x = \frac{2 \pm 2\sqrt{22}}{6}$$

$$x = \frac{1 \pm \sqrt{22}}{3}$$

Use the discriminant or a graph of a quadratic equation to determine the nature of its real solutions (zero, one, two).

The discriminant of a quadratic equation is the part of the quadratic formula that is usually inside the radical sign, $b^2 - 4ac$.

$$x = \frac{-b \pm \sqrt{b^2 - 4ac}}{2a}$$

The radical sign is NOT part of the discriminant!! Determine the value of the discriminant by substituting the values of a, b, and c from $ax^2 + bx + c = 0$.

-If the value of the discriminant is **any negative number**, then there are **two complex roots** including "i".

-If the value of the discriminant is **zero**, then there is only **1 real rational root**. This would be a double root.

-If the value of the discriminant is **any positive number that is also a perfect square**, then there are **two real rational roots**. (There are no longer any radical signs.)

-If the value of the discriminant is **any positive number that is NOT a perfect square**, then there are **two real irrational roots**. (There are still unsimplified radical signs.)

<u>Example:</u>

Find the value of the discriminant for the following equations. Then determine the number and nature of the solutions of that quadratic equation.

$$2x^2 - 5x + 6 = 0$$

$a = 2$, $b = {}^-5$, $c = 6$ so $b^2 - 4ac = ({}^-5)^2 - 4(2)(6) = 25 - 48 = {}^-23$.

Since $^-23$ is a negative number, there are **two complex roots** including "i".

$$x = \frac{5}{4} + \frac{i\sqrt{23}}{4}, \ x = \frac{5}{4} - \frac{i\sqrt{23}}{4}$$

$$3x^2 - 12x + 12 = 0$$

$a = 3$, $b = {}^-12$, $c = 12$ so $b^2 - 4ac = ({}^-12)^2 - 4(3)(12) = 144 - 144 = 0$

Since 0 is the value of the discriminant, there is only **1 real rational root**, $x = 2$

$$6x^2 - x - 2 = 0$$

$a = 6$, $b = {}^-1$, $c = {}^-2$ so $b^2 - 4ac = ({}^-1)^2 - 4(6)({}^-2) = 1 + 48 = 49$.

Since 49 is positive and is also a perfect square $(\sqrt{49}) = 7$, then there are **two real rational roots,**

$$x = \frac{2}{3}, \ x = -\frac{1}{2}$$

Try these:

1. $6x^2 - 7x - 8 = 0$

2. $10x^2 - x - 2 = 0$

3. $25x^2 - 80x + 64 = 0$

Identify graphs of relations involving quadratic inequalities.

To graph an inequality, graph the quadratic as if it was an equation; however, if the inequality has just a > or < sign, then make the curve itself dotted. Shade above the curve for > or ≥. Shade below the curve for < or ≤.

Examples:

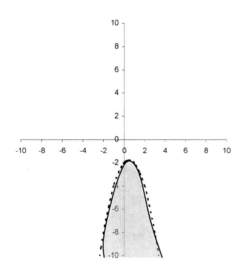

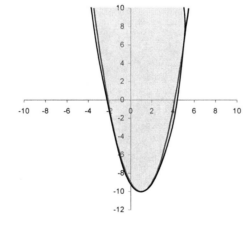

Determine the sum of terms in a progression.

The sums of terms in a progression is simply found by determining if it is an arithmetic or geometric sequence and then using the appropriate formula.

Sum of first n terms of an arithmetic sequence.

$$S_n = \frac{n}{2}(a_1 + a_n)$$

or

$$S_n = \frac{n}{2}\left[2a_1 + (n-1)d\right]$$

Sum of first n terms of a geometric sequence.

$$S_n = \frac{a_1\left(r^n - 1\right)}{r - 1}, r \neq 1$$

Sample Problem:

1. $\displaystyle\sum_{i=1}^{10}(2i + 2)$

This means find the sum of the terms beginning with the first term and ending with the 10th term of the sequence $a = 2i + 2$.

$a_1 = 2(1) + 2 = 4$

$a_{10} = 2(10) + 2 = 22$

$$S_n = \frac{n}{2}(a_1 + a_n)$$

$$S_n = \frac{10}{2}(4 + 22)$$

$$S_n = 130$$

$$S_6 = 3(-11) = -33$$

Practice problems:

1. Find the sum of the first five terms of the sequence if $a = 7$ and $d = 4$.

2. $\displaystyle\sum_{i=1}^{7}(2i - 4)$

3. $\displaystyle\sum_{i=1}^{6}{}^{-}3\left(\frac{2}{5}\right)^i$

COMPETENCY 0006   UNDERSTAND PROPERTIES OF NONLINEAR FUNCTIONS.

A rational function is given in the form $f(x) = p(x)/q(x)$. In the equation, $p(x)$ and $q(x)$ both represent polynomial functions where $q(x)$ does not equal zero. The branches of rational functions approach asymptotes. Setting the denominator equal to zero and solving will give the value(s) of the vertical asymptotes(s) since the function will be undefined at this point. If the value of $f(x)$ approaches b as the $|x|$ increases, the equation $y = b$ is a horizontal asymptote. To find the horizontal asymptote it is necessary to make a table of values for x that are to the right and left of the vertical asymptotes. The pattern for the horizontal asymptotes will become apparent as the $|x|$ increases.

If there are more than one vertical asymptotes, remember to choose numbers to the right and left of each one in order to find the horizontal asymptotes and have sufficient points to graph the function.

Example: Graph $f(x) = \dfrac{3x+1}{x-2}$.

$x - 2 = 0$ 1. Set denominator $= 0$ to find
$x = 2$ the vertical asymptote.

x	$f(x)$
3	10
10	3.875
100	3.07
1000	3.007
1	$^-4$
$^-10$	2.417
$^-100$	2.93
$^-1000$	2.99

2. Make table choosing numbers to the right and left of the vertical asymptote.

3. The pattern shows that as the $|x|$ increases $f(x)$ approaches the value 3, therefore a horizontal asymptote exists at $y = 3$

Sketch the graph.

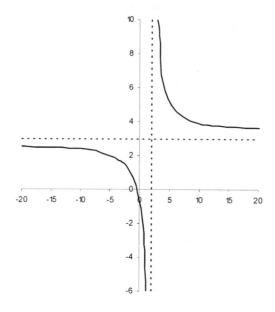

Functions defined by two or more formulas are **piecewise functions**. The formula used to evaluate piecewise functions varies depending on the value of x. The graphs of piecewise functions consist of two or more pieces, or intervals, and are often discontinuous.

Example:

$$f(x) = \begin{array}{ll} x + 1 & \text{if } x > 2 \\ x - 2 & \text{if } x \le 2 \end{array}$$

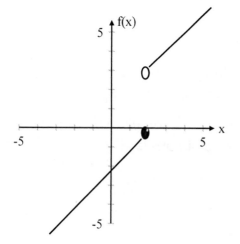

Example:

$$f(x) = \begin{array}{ll} x & \text{if } x \ge 1 \\ x^2 & \text{if } x < 1 \end{array}$$

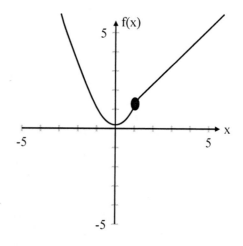

When graphing or interpreting the graph of piecewise functions it is important to note the points at the beginning and end of each interval because the graph must clearly indicate what happens at the end of each interval. Note that in the graph of Example 1, point (2, 3) is not part of the graph and is represented by an empty circle. On the other hand, point (2, 0) is part of the graph and is represented as a solid circle. Note also that the graph of Example 2 is continuous despite representing a piecewise function.

Practice Problems: Graph the following piecewise equations.

1. $f(x) = x^2$ if $x > 0$
          $= x + 4$ if $x \leq 0$

2. $f(x) = x^2 - 1$ if $x > 2$

          $= x^2 + 2$ if $x \leq 2$

Logarithmic Functions

Logarithmic functions of base a are of the basic form

$f(x) = \log_a x$, where $a > 0$ and not equal to 1.

The domain of the function, f, is (0, + inf.) and the range is (- inf., + inf.). The x-intercept of the logarithmic function is (1,0) because any number raised to the power of 0 is equal to one. The graph of the function, f, has a vertical asymptote at $x = 0$. As the value of $f(x)$ approaches negative infinity, x becomes closer and closer to 0.

Example:

Graph the function $f(x) = \log_2 (x + 1)$.

The domain of the function is all values of x such that $x + 1 > 0$. Thus, the domain of $f(x)$ is $x > -1$.

The range of $f(x)$ is (-inf., +inf.).

The vertical asymptote of $f(x)$ is the value of x that satisfies the equation $x + 1 = 0$. Thus, the vertical asymptote is $x = -1$. Note that we can find the vertical asymptote of a logarithmic function by setting the product of the logarithm (containing the variable) equal to 0.

The x-intercept of f(x) is the value of x that satisfies the equation $x + 1 = 1$ because $2^0 = 1$. Thus, the x-intercept of f(x) is (0,0). Note that we can find the x-intercept of a logarithmic function by setting the product of the logarithm equal to 1.

Finally, we find two additional values of f(x), one between the vertical asymptote and the x-intercept and the other to the right of the x-intercept. For example, f(-0.5) = -1 and f(3) = 2.

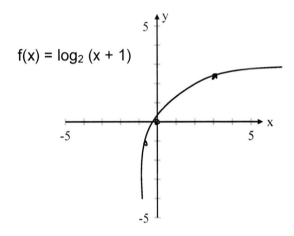

$$f(x) = \log_2 (x + 1)$$

Exponential Functions

The inverse of a logarithmic function is an exponential function. Exponential functions of base *a* take the basic form

$f(x) = a^x$, where a > 0 and not equal to 1.

The domain of the function, *f*, is (-inf., +inf.). The range is the set of all positive real numbers. If a < 1, *f* is a decreasing function and if a > 1 *f* is an increasing function. The y-intercept of f(x) is (0,1) because any base raised to the power of 0 equals 1. Finally, f(x) has a horizontal asymptote at y = 0.

Example:

Graph the function $f(x) = 2^x - 4$.

The domain of the function is the set of all real numbers and the range is y > -4. Because the base is greater than 1, the function is increasing. The y-intercept of f(x) is (0,-3). The x-intercept of f(x) is (2,0). The horizontal asymptote of f(x) is y = -4.

Finally, to construct the graph of f(x) we find two additional values for the function. For example, f(-2) = -3.75 and f(3) = 4.

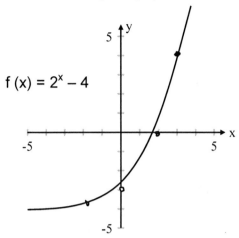

$$f(x) = 2^x - 4$$

Note that the horizontal asymptote of any exponential function of the form g(x) = a^x + b is y = b. Note also that the graph of such exponential functions is the graph of h(x) = a^x shifted b units up or down. Finally, the graph of exponential functions of the form g(x) = $a^{(x + b)}$ is the graph of h(x) = a^x shifted b units left or right.

Convert natural or common logarithmic functions into exponential functions and vice versa.

When changing common logarithms to exponential form,

$$y = \log_b x \quad \text{if and only if} \quad x = b^y$$

Natural logarithms can be changed to exponential form by using,

$$\log_e x = \ln x \quad \text{or} \quad \ln x = y \quad \text{can be written as} \quad e^y = x$$

Practice Problems:

Express in exponential form.

1. $\log_3 81 = 4$

   $x = 81 \quad b = 3 \quad y = 4$ Identify values.

   $81 = 3^4$ Rewrite in exponential form.

Solve by writing in exponential form.

2. $\log_x 125 = 3$

   $x^3 = 125$ Write in exponential form.

   $x^3 = 5^3$ Write 125 in exponential form.

   $x = 5$ Bases must be equal if exponents are equal.

Apply the properties of logarithmic or exponential functions to solve equations.

To solve logarithms or exponential functions it is necessary to use several properties.

Multiplication Property $\qquad \log_b mn = \log_b m + \log_b n$

Quotient Property $\qquad \log_b \dfrac{m}{n} = \log_b m - \log_b n$

Powers Property $\qquad \log_b n^r = r \log_b n$

Equality Property $\qquad \log_b n = \log_b m \qquad$ if and only if $n = m$.

Change of Base Formula $\qquad \log_b n = \dfrac{\log n}{\log b}$

$\qquad\qquad\qquad\qquad\qquad\qquad \log_b b^x = x$ and $b^{\log_b x} = x$

Sample problem.

Solve for x .
1. $\log_6(x-5) + \log_6 x = 2$

$\quad \log_6 x(x-5) = 2 \qquad\qquad$ Use product property.

$\quad \log_6 x^2 - 5x = 2 \qquad\qquad$ Distribute.

$\quad x^2 - 5x = 6^2 \qquad\qquad$ Write in exponential form.

$\quad x^2 - 5x - 36 = 0 \qquad\qquad$ Solve quadratic equation.

$\quad (x+4)(x-9) = 0$

$\qquad x = {}^-4 \quad x = 9$

***Be sure to check results. Remember x must be greater than zero in $\log x = y$.

Check: $\log_6(x-5) + \log_6 x = 2$

$\log_6(^-4-5) + \log_6(^-4) = 2$ Substitute the first answer $^-4$.

$\log_6(^-9) + \log_6(^-4) = 2$ This is undefined, x is less than zero.

$\log_6(9-5) + \log_6 9 = 2$ Substitute the second answer 9.

$\log_6 4 + \log_6 9 = 2$

$\log_6(4)(9) = 2$ Multiplication property.

$\log_6 36 = 2$

$6^2 = 36$ Write in exponential form.

$36 = 36$

Determine if a function is increasing or decreasing by using the first derivative in a given interval.

A function is said to be increasing if it is rising from left to right and decreasing if it is falling from left to right. Lines with positive slopes are increasing, and lines with negative slopes are decreasing. If the function in question is something other than a line, simply refer to the slopes of the tangent lines as the test for increasing or decreasing. Take the derivative of the function and plug in an x value to get the slope of the tangent line; a positive slope means the function is increasing and a negative slope means it is decreasing. If an interval for x values is given, just pick any point between the two values to substitute.

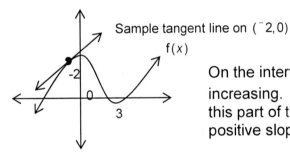

Sample tangent line on $(^-2,0)$

$f(x)$

On the interval $(^-2,0)$, $f(x)$ is increasing. The tangent lines on this part of the graph have positive slopes.

Example:

The growth of a certain bacteria is given by $f(x) = x + \dfrac{1}{x}$. Determine if the rate of growth is increasing or decreasing on the time interval $(^-1,0)$.

$f'(x) = 1 + \dfrac{^-1}{x^2}$

To test for increasing or decreasing, find the slope of the tangent line by taking the derivative.

$f'\left(\dfrac{^-1}{2}\right) = 1 + \dfrac{^-1}{(^-1/2)^2}$

Pick any point on $(^-1,0)$ and substitute into the derivative.

$f'\left(\dfrac{^-1}{2}\right) = 1 + \dfrac{^-1}{1/4}$

$= 1 - 4$

$= {}^-3$

The slope of the tangent line at $x = \dfrac{^-1}{2}$ is $^-3$. The exact value of the slope is not important. The important fact is that the slope is negative.

Find relative and absolute maxima and minima.

Substituting an x value into a function produces a corresponding y value. The coordinates of the point (x,y), where y is the largest of all the y values, is said to be a maximum point. The coordinates of the point (x,y), where y is the smallest of all the y values, is said to be a minimum point. To find these points, only a few x values must be tested. First, find all of the x values that make the derivative either zero or undefined. Substitute these values into the original function to obtain the corresponding y values. Compare the y values. The largest y value is a maximum; the smallest y value is a minimum. If the question asks for the maxima or minima on an interval, be certain to also find the y values that correspond to the numbers at either end of the interval.

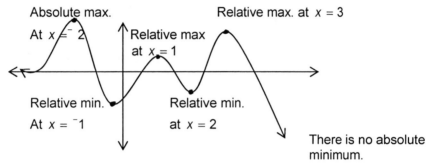

Example:

Find the maxima and minima of $f(x) = 2x^4 - 4x^2$ at the interval $(^-2,1)$.

$f'(x) = 8x^3 - 8x$ — Take the derivative first. Find all the $8x^3 - 8x = 0$ x values (critical values) that make the derivative zero or undefined.

$8x(x^2 - 1) = 0$ — In this case, there are no x values that make the derivative undefined.

$8x(x-1)(x+1) = 0$

$x = 0,\ x = 1,\ \text{or}\ x = ^- 1$ — Substitute the critical values into

$f(0) = 2(0)^4 - 4(0)^2 = 0$ — the original function. Also, plug in

$f(1) = 2(1)^4 - 4(1)^2 = ^- 2$ — the endpoint of the interval.

$f(^-1) = 2(^-1)^4 - 4(^-1)^2 = ^- 2$ — Note that 1 is a critical point and

$f(^-2) = 2(^-2)^4 - 4(^-2)^2 = 16$ — an endpoint.

The maximum is at ($^-2, 16$) and there are minima at $(1, ^-2)$ and $(^-1, ^-2)$. (0,0) is neither the maximum or minimum on ($^-2,1$) but it is still considered a relative extra point.

Find intervals on a curve where the curve is concave up or concave down.

The first derivative reveals whether a curve is rising or falling (increasing or decreasing) from the left to the right. In much the same way, the second derivative relates whether the curve is concave up or concave down. Curves which are concave up are said to "collect water;" curves which are concave down are said to "dump water." To find the intervals where a curve is concave up or concave down, follow the following steps.

1. Take the second derivative (i.e. the derivative of the first derivative).
2. Find the critical x values.
   -Set the second derivative equal to zero and solve for critical x values.
   -Find the x values that make the second derivative undefined (i.e. make the denominator of the second derivative equal to zero).
   Such values may not always exist.
3. Pick sample values which are both less than and greater than each of the critical values.
4. Substitute each of these sample values into the second derivative and determine whether the result is positive or negative.

   -If the sample value yields a positive number for the second derivative, the curve is concave up on the interval where the sample value originated.
   -If the sample value yields a negative number for the second derivative, the curve is concave down on the interval where the sample value originated.

Example:

Find the intervals where the curve is concave up and concave down for $f(x) = x^4 - 4x^3 + 16x - 16$.

$f'(x) = 4x^3 - 12x^2 + 16$ Take the second derivative.

$f''(x) = 12x^2 - 24x$

Find the critical values by setting the second derivative equal to

$12x^2 - 24x = 0$ zero.

$12x(x - 2) = 0$ There are no values that make

$x = 0$ or $x = 2$ the second derivative undefined.

Set up a number line with the critical values.

Sample values: $^-1, 1, 3$ Pick sample values in each of the 3 intervals.

$f''(^-1) = 12(^-1)^2 - 24(^-1) = 36$

If the sample value

$f''(1) = 12(1)^2 - 24(1) = {}^-12$

produces a negative number, the function is

$f''(3) = 12(3)^2 - 24(3) = 36$ concave down.

If the value produces a positive number, the curve is concave up. If

The value produces a zero, the function is linear.

Therefore when $x < 0$ the function is concave up,
       when $0 < x < 2$ the function is concave down,
       when $x > 2$ the function is concave up.

Identify points of inflection.

A point of inflection is a point where a curve changes from being concave up to concave down or vice versa. To find these points, follow the steps on the previous page for finding the intervals where a curve is concave up or concave down. A critical value is part of an inflection point if the curve is concave up on one side of the value and concave down on the other. The critical value is the x coordinate of the inflection point. To get the y coordinate, plug the critical value into the **original** function.

Example: Find the inflection points of $f(x) = 2x - \tan x$ where $\dfrac{^-\pi}{2} < x < \dfrac{\pi}{2}$.

$(x) = 2x - \tan x \qquad \dfrac{^-\pi}{2} < x < \dfrac{\pi}{2}$

Note the restriction on x.

$f'(x) = 2 - \sec^2 x$

Take the second derivative.

$f''(x) = 0 - 2 \bullet \sec x \bullet (\sec x \tan x)$

Use the Power rule.

$\quad = {}^-2 \bullet \dfrac{1}{\cos x} \bullet \dfrac{1}{\cos x} \bullet \dfrac{\sin x}{\cos x}$

The derivative of $\sec x$ is $(\sec x \tan x)$.

$f''(x) = \dfrac{^-2\sin x}{\cos^3 x}$

Find critical values by solving for the second derivative equal to zero.

$0 = \dfrac{^-2\sin x}{\cos^3 x}$

No x values on $\left(\dfrac{^-\pi}{2}, \dfrac{\pi}{2} \right)$ make the denominator zero.

$^-2\sin x = 0$
$\sin x = 0$
$\qquad x = 0$

Pick sample values on each side of the critical value $x = 0$.

Sample values: $x = \dfrac{^-\pi}{4}$ and $x = \dfrac{\pi}{4}$

$f''\left(\dfrac{^-\pi}{4} \right) = \dfrac{^-2\sin(^-\pi/4)}{\cos^3(\pi/4)} = \dfrac{^-2(^-\sqrt{2}/2)}{(\sqrt{2}/2)^3} = \dfrac{\sqrt{2}}{(\sqrt{8}/8)} = \dfrac{8\sqrt{2}}{\sqrt{8}} = \dfrac{8\sqrt{2}}{\sqrt{8}} \cdot \dfrac{\sqrt{8}}{\sqrt{8}}$

$\qquad\qquad\qquad\qquad\qquad\qquad = \dfrac{8\sqrt{16}}{8} = 4$

$f''\left(\dfrac{\pi}{4} \right) = \dfrac{^-2\sin(\pi/4)}{\cos^3(\pi/4)} = \dfrac{^-2(\sqrt{2}/2)}{(\sqrt{2}/2)^3} = \dfrac{^-\sqrt{2}}{(\sqrt{8}/8)} = \dfrac{^-8\sqrt{2}}{\sqrt{8}} = -4$

The second derivative is positive on $(0, \infty)$ and negative on $(^-\infty, 0)$. So the curve changes concavity at $x = 0$. Use the original equation to find the y value that inflection occurs at.

$f(0) = 2(0) - \tan 0 = 0 - 0 = 0$

The inflection point is $(0,0)$.

Determine and identify the intercepts and asymptotes of a given rational function.

A rational function is given in the form $f(x) = p(x)/q(x)$. In the equation, p(x) and q(x) both represent polynomial functions where q(x) does not equal zero. The branches of rational functions approach asymptotes. Setting the denominator equal to zero and solving will give the value(s) of the vertical asymptotes(s) since the function will be undefined at this point. If the value of f(x) approaches b as the $|x|$ increases, the equation $y = b$ is a horizontal asymptote. To find the horizontal asymptote it is necessary to make a table of value for x that are to the right and left of the vertical asymptotes. The pattern for the horizontal asymptotes will become apparent as the $|x|$ increases.

If there are more than one vertical asymptotes, remember to choose numbers to the right and left of each one in order to find the horizontal asymptotes and have sufficient points to graph the function.

Sample problem:

1. Graph $f(x) = \dfrac{3x + 1}{x - 2}$.

$x - 2 = 0$ $x = 2$	1. Set denominator $= 0$ to find the vertical asymptote.

x	$f(x)$	
3	10	2. Make table choosing numbers to the right and left of the vertical asymptote.
10	3.875	
100	3.07	
1000	3.007	3. The pattern shows that as the
1	‾4	$|x|$ increases f(x) approaches
‾10	2.417	the value 3, therefore a horizontal
‾100	2.93	asymptote exists at $y = 3$
‾1000	2.99	

Application of Exponential and Logarithmic Functions

In finance, the value of a sum of money with compounded interest increases at a rate proportional to the original value. We use an exponential function to determine the growth of an investment accumulating compounded interest. The formula for calculating the value of an investment after a given compounding period is

$$A(t) = A_0(1 + \frac{i}{n})^{nt}.$$

A_0 is the principle, the original value of the investment. The rate of interest is i, the time in years is t, and the number of times the interest is compounded per year is n.

We can solve the compound interest formula for any of the variables by utilizing the properties of exponents and logarithms.

1. Find the principle (A_0) that yields \$500 with an interest rate of 7.5% compounded semiannually for 20 years.

In this problem A(t) = 500, the interest rate (i) is 0.075, $n = 2$, and $t = 20$. To find the principle value, we solve for A_0.

$$A(t) = A_0(1 + \frac{i}{n})^{nt}$$

$$500 = A_0(1 + \frac{0.075}{2})^{2(20)}$$

$$A_0 = \frac{500}{1.0375^{40}} = \$114.67$$

Substitute and solve for A_0.

COMPETENCY 0007   UNDERSTAND PROPERTIES OF TRIGONOMETRIC FUNCTIONS AND IDENTITIES.

Solve equations involving circular/trigonometric functions and their inverses.

Unlike trigonometric identities that are true for all values of the defined variable, trigonometric equations are true for some, but not all, of the values of the variable. Most often trigonometric equations are solved for values between 0 and 360 degrees or 0 and 2π radians.

Some algebraic operation, such as squaring both sides of an equation, will give you extraneous answers. You must remember to check all solutions to be sure that they work.

Sample problems:

1. Solve: $\cos x = 1 - \sin x$ if $0 \le x < 360$ degrees.

$\cos^2 x = (1 - \sin x)^2$	1. square both sides
$1 - \sin^2 x = 1 - 2\sin x + \sin^2 x$	2. substitute
$0 = {}^-2\sin x + 2\sin^2 x$	3. set $=$ to 0
$0 = 2\sin x({}^-1 + \sin x)$	4. factor
$2\sin x = 0 \qquad {}^-1 + \sin x = 0$	5. set each factor $= 0$
$\sin x = 0 \qquad\quad \sin x = 1$	6. solve for $\sin x$
$x = 0$ or $180 \qquad x = 90$	7. find value of at x

The solutions appear to be 0, 90 and 180. Remember to check each solution and you will find that 180 does not give you a true equation. Therefore, the only solutions are 0 and 90 degrees.

2. Solve: $\cos^2 x = \sin^2 x$ if $0 \le x < 2\pi$

$\cos^2 x = 1 - \cos^2 x$ 1. substitute

$2\cos^2 x = 1$ 2. simplify

$\cos^2 x = \dfrac{1}{2}$ 3. divide by 2

$\sqrt{\cos^2 x} = \pm\sqrt{\dfrac{1}{2}}$ 4. take square root

$\cos x = \dfrac{\pm\sqrt{2}}{2}$ 5. rationalize denominator

$x = \dfrac{\pi}{4}, \dfrac{3\pi}{4}, \dfrac{5\pi}{4}, \dfrac{7\pi}{4}$

Prove circular/trigonometric function identities.

Given 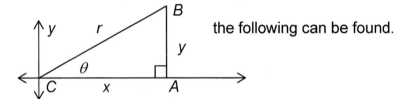 the following can be found.

Trigonometric Functions:

$\sin\theta = \dfrac{y}{r}$ $\csc\theta = \dfrac{r}{y}$

$\cos\theta = \dfrac{x}{r}$ $\sec\theta = \dfrac{r}{x}$

$\tan\theta = \dfrac{y}{x}$ $\cot\theta = \dfrac{x}{y}$

Sample problem:

1. Prove that $\sec\theta = \dfrac{1}{\cos\theta}$.

$\sec\theta = \dfrac{1}{\dfrac{x}{r}}$ Substitution definition of cosine.

$\sec\theta = \dfrac{1 \times r}{\dfrac{x}{r} \times r}$ Multiply by $\dfrac{r}{r}$.

$\sec\theta = \dfrac{r}{x}$ Substitution.

$\sec\theta = \sec\theta$ Substitute definition of $\dfrac{r}{x}$.

$\sec\theta = \dfrac{1}{\cos\theta}$ Substitute.

2. Prove that $\sin^2 + \cos^2 = 1$.

$\left(\dfrac{y}{r}\right)^2 + \left(\dfrac{x}{r}\right)^2 = 1$ Substitute definitions of sin and cos.

$\dfrac{y^2 + x^2}{r^2} = 1$ $x^2 + y^2 = r^2$ Pythagorean formula.

$\dfrac{r^2}{r^2} = 1$ Simplify.

$1 = 1$ Substitute.

$\sin^2\theta + \cos^2\theta = 1$

Practice problems: Prove each identity.

1. $\cot\theta = \dfrac{\cos\theta}{\sin\theta}$ 2. $1 + \cot^2\theta = \csc^2\theta$

Apply basic circular/trigonometric function identities.

There are two methods that may be used to prove trigonometric identities. One method is to choose one side of the equation and manipulate it until it equals the other side. The other method is to replace expressions on both sides of the equation with equivalent expressions until both sides are equal.

The Reciprocal Identities

$\sin x = \dfrac{1}{\csc x}$ $\sin x \csc x = 1$ $\csc x = \dfrac{1}{\sin x}$

$\cos x = \dfrac{1}{\sec x}$ $\cos x \sec x = 1$ $\sec x = \dfrac{1}{\cos x}$

$\tan x = \dfrac{1}{\cot x}$ $\tan x \cot x = 1$ $\cot x = \dfrac{1}{\tan x}$

$\tan x = \dfrac{\sin x}{\cos x}$ $\cot x = \dfrac{\cos x}{\sin x}$

The Pythagorean Identities

$$\sin^2 x + \cos^2 x = 1 \qquad 1 + \tan^2 x = \sec^2 x \qquad 1 + \cot^2 x = \csc^2 x$$

Sample problems:

1. Prove that $\cot x + \tan x = (\csc x)(\sec x)$.

$$\frac{\cos x}{\sin x} + \frac{\sin x}{\cos x}$$
Reciprocal identities.

$$\frac{\cos^2 x + \sin^2 x}{\sin x \cos x}$$
Common denominator.

$$\frac{1}{\sin x \cos x}$$
Pythagorean identity.

$$\frac{1}{\sin x} \times \frac{1}{\cos x}$$

$$\csc x(\sec x) = \csc x(\sec x)$$
Reciprocal identity, therefore,

$$\cot x + \tan x = \csc x(\sec x)$$

2. Prove that $\dfrac{\cos^2 \theta}{1 + 2\sin \theta + \sin^2 \theta} = \dfrac{\sec \theta - \tan \theta}{\sec \theta + \tan \theta}$.

$$\frac{1 - \sin^2 \theta}{(1 + \sin \theta)(1 + \sin \theta)} = \frac{\sec \theta - \tan \theta}{\sec \theta + \tan \theta}$$
Pythagorean identity factor denominator.

$$\frac{1 - \sin^2 \theta}{(1 + \sin \theta)(1 + \sin \theta)} = \frac{\dfrac{1}{\cos \theta} - \dfrac{\sin \theta}{\cos \theta}}{\dfrac{1}{\cos \theta} + \dfrac{\sin \theta}{\cos \theta}}$$
Reciprocal identities.

$$\frac{(1 - \sin \theta)(1 + \sin \theta)}{(1 + \sin \theta)(1 + \sin \theta)} = \frac{\dfrac{1 - \sin \theta}{\cos \theta}(\cos \theta)}{\dfrac{1 + \sin \theta}{\cos \theta}(\cos \theta)}$$
Factor $1 - \sin^2 \theta$.

Multiply by $\dfrac{\cos \theta}{\cos \theta}$.

$$\frac{1 - \sin \theta}{1 + \sin \theta} = \frac{1 - \sin \theta}{1 + \sin \theta}$$
Simplify.

$$\frac{\cos^2 \theta}{1 + 2\sin \theta + \sin^2 \theta} = \frac{\sec \theta - \tan \theta}{\sec \theta + \tan \theta}$$

The unit circle is a circle with a radius of one centered at (0,0) on the coordinate plane. Thus, any ray from the origin to a point on the circle forms an angle, t, with the positive x-axis. In addition, the ray from the origin to a point on the circle in the first quadrant forms a right triangle with a hypotenuse measuring one unit. Applying the Pythagorean theorem, $x^2 + y^2 = 1$. Because the reflections of the triangle about both the x- and y-axis are on the unit circle and $x^2 = (-x)^2$ for all x values, the formula holds for all points on the unit circle.

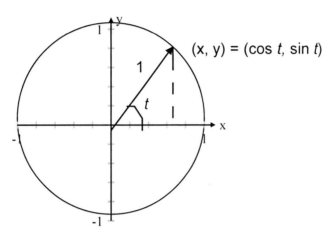

Note that because the length of the hypotenuse of the right triangles formed in the unit circle is one, the point on the unit circle that forms the triangle is (cos t, sin t). In other words, for any point on the unit circle, the x value represents the cosine of t, the y value represents the sine of t, and the ratio of the y-coordinate to the x-coordinate is the tangent of t.

The unit circle illustrates several properties of trigonometric functions. For example, applying the Pythagorean theorem to the unit circle yields the equation $\cos^2(t) + \sin^2(t) = 1$. In addition, the unit circle reveals the periodic nature of trigonometric functions. When we increase the angle t beyond 2π radians or 360 degrees, the values of x and y coordinates on the unit circle remain the same. Thus, the sine and cosine values repeat with each revolution. The unit circle also reveals the range of the sine and cosine functions. The values of sine and cosine are always between one and negative one. Finally, the unit circle shows that when the x coordinate of a point on the circle is zero, the tangent function is undefined. Thus, tangent is undefined at the angles $\dfrac{\pi}{2}, \dfrac{3\pi}{2}$ and the corresponding angles in all subsequent revolutions.

Graph trigonometric functions.

It is easiest to graph trigonometric functions when using a calculator by making a table of values.

DEGREES

	0	30	45	60	90	120	135	150	180	210	225	240	270	300	315	330	360
sin	0	.5	.71	.87	1	.87	.71	.5	0	-.5	-.71	-.87	-1	-.87	-.71	-.5	0
cos	1	.87	.71	.5	0	-.5	-.71	-.87	-1	-.87	-.71	-.5	0	.5	.71	.87	1
tan	0	.57	1	1.7	--	-1.7	-1	-.57	0	.57	1	1.7	--	-1.7	-1	-.57	0

$$0 \quad \frac{\pi}{6} \quad \frac{\pi}{4} \quad \frac{\pi}{3} \quad \frac{\pi}{2} \quad \frac{2\pi}{3} \quad \frac{3\pi}{4} \quad \frac{5\pi}{6} \quad \pi \quad \frac{7\pi}{6} \quad \frac{5\pi}{4} \quad \frac{4\pi}{3} \quad \frac{3\pi}{2} \quad \frac{5\pi}{3} \quad \frac{7\pi}{4} \quad \frac{11\pi}{6} \quad 2\pi$$

RADIANS

Remember the graph always ranges from +1 to ⁻1 for sine and cosine functions unless noted as the coefficient of the function in the equation. For example, $y = 3\cos x$ has an amplitude of 3 units from the center line (0). Its maximum and minimum points would be at +3 and ⁻3.

Tangent is not defined at the values 90 and 270 degrees or $\frac{\pi}{2}$ and $\frac{3\pi}{2}$. Therefore, vertical asymptotes are drawn at those values.

The inverse functions can be graphed in the same manner using a calculator to create a table of values.

Solve a triangle, given appropriate parts.

In order to solve a right triangle using trigonometric functions it is helpful to identify the given parts and label them. Usually more than one trigonometric function may be appropriately applied.

Some items to know about right triangles:

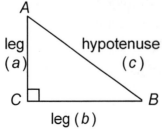

Given angle A, the side labeled leg (a) Is adjacent angle A. And the side(b) is opposite to angle A

Sample problem:

1. Find the missing side.

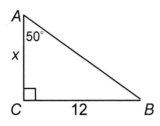

1. Identify the known values. Angle $A = 50$ degrees. The side opposite the given angle is 12. The missing side is the adjacent leg.
2. The information suggests the use of the tangent function

$$\tan A = \frac{\text{opposite}}{\text{adjacent}}$$

3. Write the function.

$$\tan 50 = \frac{12}{x}$$

4. Substitute.

$$1.192 = \frac{12}{x}$$

5. Solve.

$$x(1.192) = 12$$

$$x = 10.069$$

Remember that since angle A and angle B are complimentary, then angle $B = 90 - 50$ or 40 degrees.

Using this information we could have solved for the same side only this time it is the leg opposite from angle B.

$$\tan B = \frac{\text{opposite}}{\text{adjacent}}$$

1. Write the formula.

$$\tan 40 = \frac{x}{12}$$

2. Substitute.

$$12(.839) = x$$

$$10.069 \approx x$$

3. Solve.

Now that the two sides of the triangle are known, the third side can be found using the Pythagorean Theorem.

Apply the law of cosines.

Definition: For any triangle ABC, when given two sides and the included angle, the other side can be found using one of the formulas below:

$$a^2 = b^2 + c^2 - (2bc)\cos A$$
$$b^2 = a^2 + c^2 - (2ac)\cos B$$
$$c^2 = a^2 + b^2 - (2ab)\cos C$$

Similarly, when given three sides of a triangle, the included angles can be found using the derivation:

$$\cos A = \frac{b^2 + c^2 - a^2}{2bc}$$
$$\cos B = \frac{a^2 + c^2 - b^2}{2ac}$$
$$\cos C = \frac{a^2 + b^2 - c^2}{2ab}$$

Sample problem:

1. Solve triangle ABC, if angle $B = 87.5°$, $a = 12.3$, and $c = 23.2$. (Compute to the nearest tenth).

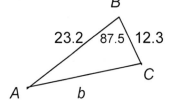

1. Draw and label a sketch.

Find side b.

$$b^2 = a^2 + c^2 - (2ac)\cos B$$

2. Write the formula.

$$b^2 = (12.3)^2 + (23.2)^2 - 2(12.3)(23.2)(\cos 87.5)$$

3. Substitute.

$$b^2 = 664.636$$
$$b = 25.8 \text{ (rounded)}$$

4. Solve.

Use the law of sines to find angle A.

$$\frac{\sin A}{a} = \frac{\sin B}{b}$$

1. Write formula.

$$\frac{\sin A}{12.3} = \frac{\sin 87.5}{25.8} = \frac{12.29}{25.8}$$

2. Substitute.

$$\sin A = 0.47629$$

3. Solve.

Angle $A = 28.4$

Therefore, angle $C = 180 - (87.5 + 28.4)$
$$= 64.1$$

2. Solve triangle ABC if $a = 15$, $b = 21$, and $c = 18$. (Round to the nearest tenth).

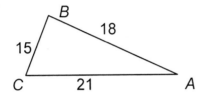

1. Draw and label sketch.

Find angle A.

$$\cos A = \frac{b^2 + c^2 - a^2}{2bc}$$

2. Write formula.

$$\cos A = \frac{21^2 + 18^2 - 15^2}{2(21)(18)}$$

3. Substitute.

$$\cos A = 0.714$$

4. Solve.

Angle $A = 44.4$

Find angle B.

$$\cos B = \frac{a^2 + c^2 - b^2}{2ac}$$

5. Write formula.

$$\cos B = \frac{15^2 + 18^2 - 21^2}{2(15)(18)}$$

6. Substitute.

$$\cos B = 0.2$$

7. Solve.

Angle $B = 78.5$

Therefore, angle $C = 180 - (44.4 + 78.5)$
$$= 57.1$$

Apply the law of sines.

Definition: For any triangle ABC, where a, b, and c are the lengths of the sides opposite angles A, B, and C respectively.

$$\frac{\sin A}{a} = \frac{\sin B}{b} = \frac{\sin C}{c}$$

Sample problem:

1. An inlet is 140 feet wide. The lines of sight from each bank to an approaching ship are 79 degrees and 58 degrees. What are the distances from each bank to the ship?

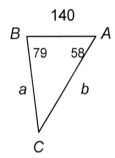

1. Draw and label a sketch.
2. The missing angle is $180 - (79 + 58) = 43°$

$$\frac{\sin A}{a} = \frac{\sin B}{b} = \frac{\sin C}{c}$$

3. Write formula.

Side opposite 79 degree angle:

$$\frac{\sin 79}{b} = \frac{\sin 43}{140}$$

4. Substitute.

$$b = \frac{140(.7826)}{.6820}$$

5. Solve.

$$b \approx 201.501 \text{ feet}$$

Side opposite 58 degree angle:

$$\frac{\sin 58}{a} = \frac{\sin 43}{140}$$

6. Substitute.

$$a = \frac{140(.8480)}{.6820}$$

7. Solve.

$$a \approx 174.076 \text{ feet}$$

Mathematicians use trigonometric functions to model and solve problems involving naturally occurring periodic phenomena. Examples of periodic phenomena found in nature include all forms of radiation (ultraviolet rays, visible light, microwaves, etc.), sound waves and pendulums. Additionally, trigonometric functions often approximate fluctuations in temperature, employment and consumer behavior (business models). The following are examples of the use of the sine function to model problems.

Example 1:

Consider the average monthly temperatures for a hypothetical location.

Month	Avg. Temp. (F)
Jan	40
March	48
May	65
July	81
Sept	80
Nov	60

Note that the graph of the average temperatures resembles the graph of a trigonometric function with a period of one year. We can use the periodic nature of seasonal temperature fluctuation to predict weather patterns.

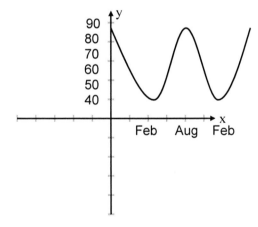

The trigonometric functions sine, cosine, and tangent are periodic functions. The values of periodic functions repeat on regular intervals. Period, amplitude, and phase shift are key properties of periodic functions that can be determined by observation of the graph.

The **period** of a function is the smallest domain containing the complete cycle of the function. For example, the period of a sine or cosine function is the distance between the peaks of the graph.

The **amplitude** of a function is half the distance between the maximum and minimum values of the function.

Phase shift is the amount of horizontal displacement of a function from its original position.

Both sine and cosine graphs are periodic waves. Below is a generic sine/cosine graph with the period and amplitude labeled.

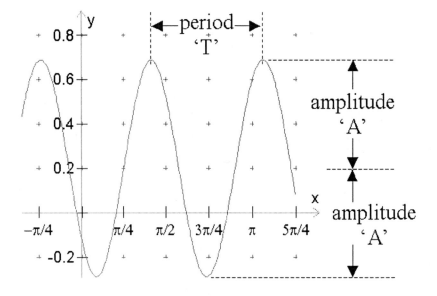

Properties of the graphs of basic trigonometric functions.

Function	Period	Amplitude
y = sin x	2π radians	1
y = cos x	2π radians	1
y = tan x	π radians	undefined

Below are the graphs of the basic trigonometric functions, (a) y = sin x; (b) y = cos x; and (c) y= tan x.

A)   **B)**   **C)**

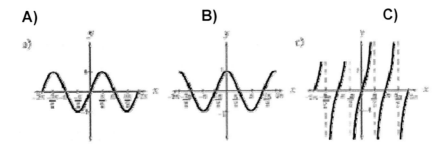

Note that the phase shift of trigonometric graphs is the horizontal distance displacement of the curve from these basic functions.

COMPETENCY 0008   UNDERSTAND PRINCIPLES AND APPLICATIONS OF CALCULUS.

A sequence is a list of numbers that follow a specific pattern. For example, the following list of numbers represents a sequence defined by the formula $\dfrac{n}{n+1}$.

$$\frac{1}{2}, \frac{2}{3}, \frac{3}{4}, \frac{4}{5}, \frac{5}{6}, \dots \frac{99}{100}, \dots$$

We say the limit of the sequence is 1, because as n approaches infinity, the sequence approaches 1. Thus, a limit is the upper or lower boundary of a sequence; the value that the sequence will never pass.

A series is a sum of numbers. Convergent series possess a numerical limit.

For example, the sum of the series $\dfrac{1}{2^n}$ starting with n = 1, has a limit of 1. We represent the series as follows:

$$\sum_{n=1}^{\infty} \frac{1}{2^n} = \frac{1}{2} + \frac{1}{4} + \frac{1}{8} + \frac{1}{16} + \frac{1}{32} + \dots$$

The symbol Σ (sigma) represents summation and the numbers to the right of the sigma symbol are the beginning and end values of the series. Note, that no matter how many values we add in the series, the total sum approaches, but never reaches 1.

Solve problems using the limit theorems concerning sums, products, and quotients of functions.

The limit of a function is the y value that the graph approaches as the x values approach a certain number. To find a limit there are two points to remember.

1. Factor the expression completely and cancel all common factors in fractions.

2. Substitute the number to which the variable is approaching. In most cases this produces the value of the limit.

If the variable in the limit is approaching ∞, factor and simplify first; then examine the result. If the result does not involve a fraction with the variable in the denominator, the limit is usually also equal to ∞. If the variable is in the denominator of the fraction, the denominator is getting larger which makes the entire fraction smaller. In other words the limit is zero.

Examples:

1. $\lim\limits_{x \to {}^-3} \dfrac{x^2+5x+6}{x+3} + 4x$ 　　Factor the numerator.

$\lim\limits_{x \to {}^-3} \dfrac{(x+3)(x+2)}{(x+3)} + 4x$ 　　Cancel the common factors.

$\lim\limits_{x \to {}^-3} (x+2) + 4x$ 　　Substitute ${}^-3$ for x.

$({}^-3+2)+4({}^-3)$ 　　Simplify.

$\quad {}^-1 + {}^-12$

$\quad\quad {}^-13$

2. $\lim\limits_{x \to \infty} \dfrac{2x^2}{x^5}$ 　　Cancel the common factors.

$\lim\limits_{x \to \infty} \dfrac{2}{x^3}$

Since the denominator is getting larger, the entire fraction is getting smaller. The fraction is getting close to zero.

$\dfrac{2}{\infty^3}$

Practice problems:

1. $\lim\limits_{x \to \pi} 5x^2 + \sin x$ 　　　　2. $\lim\limits_{x \to {}^-4} \dfrac{x^2+9x+20}{x+4}$

Find the derivatives of algebraic, trigonometric, exponential, and logarithmic functions.

Derivatives of algebraic functions.

A. Derivative of a constant--for any constant, the derivative is always zero.

B. Derivative of a variable--the derivative of a variable (i.e. x) is one.

C. Derivative of a variable raised to a power--for variable expressions with rational exponents (i.e. $3x^2$) multiply the coefficient (3) by the exponent (2) then subtract one (1) from the exponent to arrive at the derivative $\left(3x^2 = 6x\right)$.

Example:

1. $y = 5x^4$ Take the derivative.

$$\frac{dy}{dx} = (5)(4)x^{4-1}$$

     Multiply the coefficient by the exponent and subtract 1 from the exponent.

$$\frac{dy}{dx} = 20x^3$$ Simplify.

2. $y = \dfrac{1}{4x^3}$ Rewrite using negative exponent.

$$y = \frac{1}{4}x^{-3}$$ Take the derivative.

$$\frac{dy}{dx} = \left(\frac{1}{4} \bullet {}^{-}3\right)x^{-3-1}$$

$$\frac{dy}{dx} = \frac{{}^{-}3}{4}x^{-4} = \frac{{}^{-}3}{4x^4}$$ Simplify.

3. $y = 3\sqrt{x^5}$ Rewrite using $\sqrt[z]{x^n} = x^{n/z}$.

$y = 3x^{5/2}$ Take the derivative.

$$\frac{dy}{dx} = (3)\left(\frac{5}{2}\right)x^{5/2-1}$$

$$\frac{dy}{dx} = \left(\frac{15}{2}\right)x^{3/2}$$ Simplify.

$$\frac{dy}{dx} = 7.5\sqrt{x^3} = 7.5x\sqrt{x}$$

Derivatives of trigonometric functions.

A. $\sin x$ --the derivative of the sine of x is simply the cosine of x.

B. $\cos x$ --the derivative of the cosine of x is negative one ($^-1$) times the sine of x.

C. $\tan x$ --the derivative of the tangent of x is the square of the secant of x.

If object of the trig. function is an expression other than x, follow the above rules substituting the expression for x. The only additional step is to multiply the result by the derivative of the expression.

Examples:

1. $y = \pi \sin x$ Carry the coefficient (π) throughout the problem.

   $\dfrac{dy}{dx} = \pi \cos x$

2. $y = \dfrac{2}{3} \cos x$

   Do not forget to multiply the coefficient by negative one when taking the derivative of a cosine function.

   $\dfrac{dy}{dx} = \dfrac{^-2}{3} \sin x$

3. $y = 4\tan\left(5x^3\right)$

   $\dfrac{dy}{dx} = 4\sec^2\left(5x^3\right)\left(5 \bullet 3x^{3-1}\right)$

   The derivative of $\tan x$ is $\sec^2 x$.

   $\dfrac{dy}{dx} = 4\sec^2\left(5x^3\right)\left(15x^2\right)$

   Carry the $\left(5x^3\right)$ term throughout the problem.

   $\dfrac{dy}{dx} = 4 \bullet 15x^2 \sec^2\left(5x^3\right)$

   Multiply $4\sec^2\left(5x^3\right)$ by the derivative of $5x^3$.

   $\dfrac{dy}{dx} = 60x^2 \sec^2\left(5x^3\right)$

   Rearrange the terms and simplify.

Derivatives of exponential functions.

$f(x) = e^x$ is an exponential function. The derivative of e^x is exactly the same thing $\rightarrow e^x$. If instead of x, the exponent on e is an expression, the derivative is the same e raised to the algebraic exponent multiplied by the derivative of the algebraic expression.

If a base other than e is used, the derivative is the natural log of the base times the original exponential function times the derivative of the exponent.

Examples:

1. $y = e^x$

$$\frac{dy}{dx} = e^x$$

2. $y = e^{3x}$

$$\frac{dy}{dx} = e^{3x} \bullet 3 = 3e^{3x}$$

Multiply e^{3x} by the derivative of $3x$ which is 3.

$$\frac{dy}{dx} = 3e^{3x}$$

Rearrange the terms.

3. $y = \dfrac{5}{e^{\sin x}}$

$y = 5e^{-\sin x}$

Rewrite using negative exponents

$$\frac{dy}{dx} = 5e^{-\sin x} \bullet \left(^-\cos x\right)$$

Multiply $5e^{-\sin x}$ by the derivative of $^-\sin x$ which is $^-\cos x$.

$$\frac{dy}{dx} = \frac{^-5\cos x}{e^{\sin x}}$$

Use the definition of negative exponents to simplify.

4. $y = {}^-2 \bullet \ln 3^{4x}$

$$\frac{dy}{dx} = {}^-2 \bullet (\ln 3)\left(3^{4x}\right)(4)$$

The natural log of the base is ln3.

The derivative of $4x$ is 4.

$$\frac{dy}{dx} = {}^-8 \bullet 3^{4x} \ln 3$$

Rearrange terms to simplify.

Derivatives of logarithmic functions.

The most common logarithmic function on the Exam is the natural logarithmic function ($\ln x$). The derivative of $\ln x$ is simply $1/x$. If x is replaced by an algebraic expression, the derivative is the fraction one divided by the expression multiplied by the derivative of the expression.

For all other logarithmic functions, the derivative is 1 over the argument of the logarithm multiplied by 1 over the natural logarithm (ln) of the base multiplied by the derivative of the argument. Examples:

1. $y = \ln x$

    $\dfrac{dy}{dx} = \dfrac{1}{x}$

2. $y = 3\ln\left(x^{-2}\right)$

    $\dfrac{dy}{dx} = 3 \bullet \dfrac{1}{x^{-2}} \bullet \left(^{-}2x^{-2-1}\right)$

    Multiply one over the argument $\left(x^{-2}\right)$ by the derivative of x^{-2} which is $^{-}2x^{-2-1}$.

    $\dfrac{dy}{dx} = 3 \bullet x^2 \bullet \left(^{-}2x^{-3}\right)$

    $\dfrac{dy}{dx} = \dfrac{^{-}6x^2}{x^3}$

    Simplify using the definition of negative exponents.

    $\dfrac{dy}{dx} = \dfrac{^{-}6}{x}$

    Cancel common factors to simplify.

3. $y = \log_5(\tan x)$

    $\dfrac{dy}{dx} = \dfrac{1}{\tan x} \bullet \dfrac{1}{\ln 5} \bullet (\sec^2 x)$

    The derivative of $\tan x$ is $\sec^2 x$.

    $\dfrac{dy}{dx} = \dfrac{\sec^2 x}{(\tan x)(\ln 5)}$

Identify and apply definitions of the derivative of a function.

The derivative of a function has two basic interpretations.

    I.   Instantaneous rate of change
    II.   Slope of a tangent line at a given point

If a question asks for the rate of change of a function, take the derivative to find the equation for the rate of change. Then plug in for the variable to find the instantaneous rate of change.

The following is a list summarizing some of the more common quantities referred to in rate of change problems.

area	height	profit
decay	population growth	sales
distance	position	temperature
frequency	pressure	volume

Pick a point, say $x = {}^-3$, on the graph of a function. Draw a tangent line at that point. Find the derivative of the function and plug in $x = {}^-3$. The result will be the slope of the tangent line.

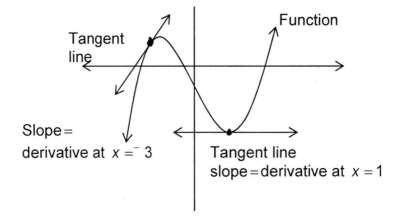

Tangent line

Function

Slope = derivative at $x = {}^-3$

Tangent line slope = derivative at $x = 1$

Use the derivative to find the slope of a curve at a point.

To find the slope of a curve at a point, there are two steps to follow.

     1.   Take the derivative of the function.
     2.   Plug in the value to find the slope.

If plugging into the derivative yields a value of zero, the tangent line is horizontal at that point.

If plugging into the derivative produces a fraction with zero in the denominator, the tangent line at this point has an undefined slope and is thus a vertical line.

Examples:

1. Find the slope of the tangent line for the given function at the given point.

$$y = \frac{1}{x-2} \text{ at } (3,1)$$

$$y = (x-2)^{-1} \qquad \text{Rewrite using negative exponents.}$$

$$\frac{dy}{dx} = {}^- 1(x-2)^{-1-1}(1)$$

Use the Chain rule. The derivative of $(x-2)$ is 1.

$$\frac{dy}{dx} = {}^- 1(x-2)^{-2}$$

$$\frac{dy}{dx}\bigg|_{x=3} = {}^- 1(3-2)^{-2} \qquad \text{Evaluate at } x = 3.$$

$$\frac{dy}{dx}\bigg|_{x=3} = {}^- 1$$

The slope of the tangent line is $^-1$ at $x = 3$.

2. Find the points where the tangent to the curve $f(x) = 2x^2 + 3x$ is parallel to the line $y = 11x - 5$.

$f'(x) = 2 \bullet 2x^{2-1} + 3$ Take the derivative of $f(x)$ to get the slope of a tangent line.

$f'(x) = 4x + 3$

$4x + 3 = 11$ Set the slope expression $(4x + 3)$ equal to the slope of $y = 11x - 5$.

$x = 2$ Solve for the x value of the point.

$f(2) = 2(2)^2 + 3(2)$ The y value is 14.

$f(2) = 14$ So (2,14) is the point on $f(x)$ where the tangent line is parallel to $y = 11x - 5$.

Find the equation of a tangent line or a normal line at a point on a curve.

To write an equation of **a tangent line** at a point, two things are needed.

A point--the problem will usually provide a point, (x, y). If the problem only gives an x value, plug the value into the original function to get the y coordinate.

The slope--to find the slope, take the derivative of the original function. Then plug in the x value of the point to get the slope.

After obtaining a point and a slope, use the Point-Slope form for the equation of a line:

$$(y - y_1) = m(x - x_1)$$

where m is the slope and (x_1, y_1) is the point.

Example:

Find the equation of the tangent line to $f(x) = 2e^{x^2}$ at $x = {}^-1$.

$f({}^-1) = 2e^{({}^-1)^2}$ Plug in the x coordinate to obtain the y coordinate.

$\quad\quad = 2e^1$ The point is $({}^-1, 2e)$.

$f'(x) = 2e^{x^2} \bullet (2x)$

$f'({}^-1) = 2e^{({}^-1)^2} \bullet (2 \bullet {}^-1)$

$f'({}^-1) = 2e^1({}^-2)$

$f'({}^-1) = {}^-4e$ The slope at $x = {}^-1$ is ${}^-4e$.

$(y - 2e) = {}^-4e(x - {}^-1)$ Plug in the point $({}^-1, 2e)$ and the slope $m = {}^-4e$. Use the point slope form of a line.

$y = {}^-4ex - 4e + 2e$ line.

$y = {}^-4ex - 2e$ Simplify to obtain the equation for the tangent line.

A **normal line** is a line which is perpendicular to a tangent line at a given point. Perpendicular lines have slopes which are negative reciprocals of each other. To find the equation of a normal line, first get the slope of the tangent line at the point. Find the negative reciprocal of this slope. Next, use the new slope and the point on the curve, both the x_1 and y_1 coordinates, and substitute into the Point-Slope form of the equation for a line:

$$(y - y_1) = slope \bullet (x - x_1)$$

Example:

1. Find the equation of the normal line to the tangent to the curve $y = (x^2 - 1)(x - 3)$ at $x = {}^-2$.

$f(-2) = (({}^-2)^2 - 1)({}^-2 - 3)$	First find the y coordinate of the point on the curve. Here,
$f(-2) = {}^-15$	$y = {}^-15$ when $x = {}^-2$.
$y = x^3 - 3x^2 - x + 3$	Before taking the derivative, multiply the expression first. The derivative of a sum is easier to find than the derivative of a product.
$y' = 3x^2 - 6x - 1$	Take the derivative to find the slope of the tangent line.
$y'_{x={}^-2} = 3({}^-2)^2 - 6({}^-2) - 1$	
$y'_{x={}^-2} = 23$	
slope of normal $= \dfrac{{}^-1}{23}$	For the slope of the normal line, take the negative reciprocal of the tangent line's slope.
$(y - {}^-15) = \dfrac{{}^-1}{23}(x - {}^-2)$	Plug (x_1, y_1) into the point-slope equation.
$(y + 15) = \dfrac{{}^-1}{23}(x + 2)$	

$$y = -\frac{1}{23}x + \frac{2}{23} - 15 = -\frac{1}{23}x - 14\frac{21}{23}$$

The **definite integral** is defined as the limit approached by the *n*th upper and lower Riemann sums as $n \to \infty$. The **Riemann sum** is the sum of the areas of all rectangles approximating the area under the graph of a function.

Example: For $f(x) = x^2$, find the values of the Riemann Sums over the interval [0, 1] using *n* subintervals of equal width evaluated at the midpoint of each subinterval. Find the limit of the Riemann Sums.

$$\int_0^1 x^2 dx$$

Let *n* be a positive integer and let $\displaystyle\lim_{n \to \infty} \frac{n(n+1)(2n+1)}{6n^3} = \frac{1}{3}$.

Take the interval [0,1] and subdivide it into *n* subintervals each of length $\dfrac{1}{n}$.

$\Delta x = 1/n$

Let $a_i = \dfrac{i}{n}$; the endpoints of the ith subinterval are

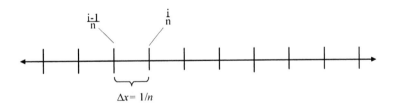

$\Delta x = 1/n$

Let $x_i = \dfrac{i}{n}$ be the right-hand endpoint. Draw a line of length

$f(x_i) = \left[\dfrac{i}{n}\right]^2$ at the right-hand endpoint.

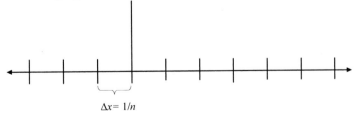

$\Delta x = 1/n$

Draw a rectangle.

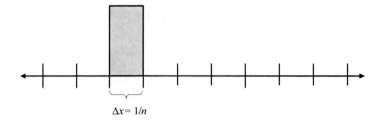

$$\Delta x = 1/n$$

The area of this rectangle is $f(x)\Delta x = \left[\dfrac{i}{n}\right]^2 \dfrac{1}{n} = \dfrac{i^2}{n^3}$.

Now draw all 9 rectangles.

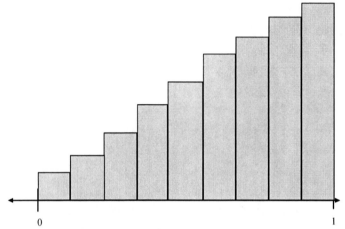

The sum of the area of these rectangles is:

$$\sum_{i=1}^{n} f(x_i)\Delta x = \sum_{i=1}^{n} \frac{i^2}{n^3} = \frac{1}{n^3}\sum_{i=1}^{n} i^2 = \frac{n(n+1)(2n+1)}{6n^3}.$$

Finally, to evaluate the integral, we take the limit

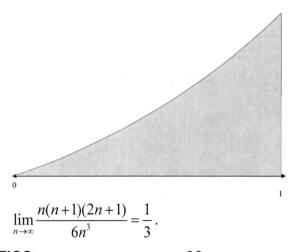

$$\lim_{n\to\infty} \frac{n(n+1)(2n+1)}{6n^3} = \frac{1}{3}.$$

The **difference quotient** is the average rate of change over an interval. For a function f, the **difference quotient** is represented by the formula:

$$\frac{f(x+h)-f(x)}{h}.$$

This formula computes the slope of the secant line through two points on the graph of f. These are the points with x-coordinates x and $x+h$.

Example: Find the difference quotient for the function $f(x) = 2x^2 + 3x - 5$.

$$\frac{f(x+h)-f(x)}{h} = \frac{2(x+h)^2 + 3(x+h) - 5 - (2x^2 + 3x - 5)}{h}$$

$$= \frac{2(x^2 + 2hx + h^2) + 3x + 3h - 5 - 2x^2 - 3x + 5}{h}$$

$$= \frac{2x^2 + 4hx + 2h^2 + 3x + 3h - 5 - 2x^2 - 3x + 5}{h}$$

$$= \frac{4hx + 2h^2 + 3h}{h}$$

$$= 4x + 2h + 3$$

The **derivative** is the slope of a tangent line to a graph $f(x)$, and is usually denoted $f'(x)$. This is also referred to as the instantaneous rate of change.

The derivative of $f(x)$ at $x = a$ is given by taking the limit of the average rates of change (computed by the difference quotient) as h approaches 0.

$$f'(a) = \lim_{h \to 0} \frac{f(a+h) - f(a)}{h}$$

Example: Suppose a company's annual profit (in millions of dollars) is represented by the above function $f(x) = 2x^2 + 3x - 5$ and x represents the number of years in the interval. Compute the rate at which the annual profit was changing over a period of 2 years.

$$f'(a) = \lim_{h \to 0} \frac{f(a+h) - f(a)}{h}$$

$$= f'(2) = \lim_{h \to 0} \frac{f(2+h) - f(2)}{h}$$

Using the difference quotient we computed above, $4x + 2h + 3$, we get

$$f'(2) = \lim_{h \to 0}(4(2) + 2h + 3)$$

$$= 8 + 3$$

$$= 11.$$

We have, therefore, determined that the annual profit for the company has increased at the average rate of $11 million per year over the two-year period.

Solve problems using instantaneous rates of change and related rates of change, such as growth and decay.

Finding the rate of change of one quantity (for example distance, volume, etc.) with respect to time it is often referred to as a rate of change problem. To find an instantaneous rate of change of a particular quantity, write a function in terms of time for that quantity; then take the derivative of the function. Substitute in the values at which the instantaneous rate of change is sought.

Functions which are in terms of more than one variable may be used to find related rates of change. These functions are often not written in terms of time. To find a related rate of change, follow these steps.

1. Write an equation which relates all the quantities referred to in the problem.
2. Take the derivative of both sides of the equation with respect to time. Follow the same steps as used in implicit differentiation. This means take the derivative of each part of the equation remembering to multiply each term by the derivative of the variable involved with respect to time. For example, if a term includes the variable v for volume, take the derivative of the term remembering to multiply by dv/dt for the derivative of volume with respect to time. dv/dt is the rate of change of the volume.
3. Substitute the known rates of change and quantities, and solve for the desired rate of change.

Example:

1. What is the instantaneous rate of change of the area of a circle where the radius is 3 cm?

$A(r) = \pi r^2$ Write an equation for area.

$A'(r) = 2\pi r$ Take the derivative to find the rate of change.

$A'(3) = 2\pi(3) = 6\pi$ Substitute in $r = 3$ to arrive at the instantaneous rate of change.

Solve distance, area, and volume problems using integration.

To find the **distance** function, take the antiderivative of the velocity function. And to find the velocity function, find the antiderivative of the acceleration function. Use the information in the problem to solve for the constants that result from taking the antiderivatives.

Example:

A particle moves along the x axis with acceleration $a(t) = 6t - 6$. The initial velocity is 0 m/sec and the initial position is 8 cm to the right of the origin. Find the velocity and position functions.

$v(0) = 0$

$s(0) = 8$

Interpret the given information.

$a(t) = 6t - 6$

Put in the coefficients needed to take the antiderivative.

$a(t) = 6 \bullet \dfrac{1}{2} \bullet 2t - 6$

$v(t) = \dfrac{6}{2}t^2 - 6t + c$

Take the antiderivative of $a^{(t)}$ to get $v^{(t)}$.

$v(0) = 3(0)^3 - 6(0) + c = 0$

Use $v(0) = 0$ to solve for c.

$0 - 0 + c = 0$

$c = 0$

$c = 0$

$v(t) = 3t^2 - 6t + 0$

Rewrite $v(t)$ using $c = 0$.

$v(t) = 3t^2 - 6\dfrac{1}{2} \bullet 2t$

Put in the coefficients needed to take the antiderivative.

$s(t) = t^3 - \dfrac{6}{2}t^2 + c$

Take the antiderivative of $v^{(t)}$ to get $s^{(t)} \rightarrow$ the distance function.

$s(0) = 0^3 - 3(0)^2 + c = 8$

Use $s(0) = 8$ to solve for c.

$\qquad\qquad c = 8$

$s(t) = t^3 - 3t^2 + 8$

Taking the integral of a function and evaluating it from one x value to another provides the total **area under the curve** (i.e. between the curve and the x axis). Remember, though, that regions above the x axis have "positive" area and regions below the x axis have "negative" area. You must account for these positive and negative values when finding the area under curves. Follow these steps.

1. Determine the x values that will serve as the left and right boundaries of the region.
2. Find all x values between the boundaries that are either solutions to the function or are values which are not in the domain of the function. These numbers are the interval numbers.
3. Integrate the function.
4. Evaluate the integral once for each of the intervals using the boundary numbers.
5. If any of the intervals evaluates to a negative number, make it positive (the negative simply tells you that the region is below the x axis).
6. Add the value of each integral to arrive at the area under the curve.

Example:

Find the area under the following function on the given intervals.
$f(x) = \sin x$; $(0, 2\pi)$

$\sin x = 0$ Find any roots to f(x) on $(0, 2\pi)$.

$\quad x = \pi$

$(0, \pi) \quad (\pi, 2\pi)$ Determine the intervals using the boundary numbers and the roots.

$\int \sin x \, dx = ^-\cos x$ Integrate f(x). We can ignore the constant c because we have numbers to use to evaluate the integral.

$^-\cos x \Big]_{x=0}^{x=\pi} = ^-\cos \pi - (^-\cos 0)$

$^-\cos x \Big]_{x=0}^{x=\pi} = ^-(-1) + (1) = 2$

$^-\cos x \Big]_{x=\pi}^{x=2\pi} = ^-\cos 2\pi - (^-\cos \pi)$

$^-\cos x \Big]_{x=\pi}^{x=2\pi} = ^-1 + (^-1) = ^-2$ The $^-2$ means that for $(\pi, 2\pi)$, the region is below the x axis, but the area is still 2.

$\text{Area} = 2 + 2 = 4$ Add the 2 integrals together to get the area.

Finding the **area between two curves** is much the same as finding the area under one curve. But instead of finding the roots of the functions, you need to find the x values which produce the same number from both functions (set the functions equal and solve). Use these numbers and the given boundaries to write the intervals. On each interval you must pick sample values to determine which function is "on top" of the other. Find the integral of each function. For each interval, subtract the "bottom" integral from the "top" integral. Use the interval numbers to evaluate each of these differences. Add the evaluated integrals to get the total area between the curves.

Example:

Find the area of the regions bounded by the two functions on the indicated intervals.

$f(x) = x + 2$ and $g(x) = x^2$ $\left[^-2, 3 \right]$ Set the functions

equal and

$x + 2 = x^2$ solve.
$0 = (x - 2)(x + 1)$

$x = 2$ or $x = ^-1$ Use the solutions and the

$(^-2, ^-1)$ $(^-1, 2)$ $(2, 3)$ boundary numbers to write

the intervals.

$f(^-3/2) = \left(\dfrac{^-3}{2} \right) + 2 = \dfrac{1}{2}$ Pick sample values on the

integral and evaluate each
function as that number.

$g(^-3/2) = \left(\dfrac{^-3}{2} \right)^2 = \dfrac{9}{4}$ $g(x)$ is "on top" on $\left[^-2, ^-1 \right]$.

$f(0) = 2$ $f(x)$ is "on top" on $\left[^-1, 2 \right]$.

$g(0) = 0$

$f(5/2) = \dfrac{5}{2} + 2 = \dfrac{9}{2}$ $g(x)$ is "on top" on $[2, 3]$.

$g(5/2) = \left(\dfrac{5}{2} \right)^2 = \dfrac{25}{4}$

$\int f(x)dx = \int (x + 2)dx$

$$\int f(x)dx = \int xdx + 2\int dx$$

$$\int f(x)dx = \frac{1}{1+1}x^{1+1} + 2x$$

$$\int f(x)dx = \frac{1}{2}x^2 + 2x$$

$$\int g(x)dx = \int x^2 dx$$

$$\int g(x)dx = \frac{1}{2+1}x^{2+1} = \frac{1}{3}x^3$$

Area $1 = \int g(x)dx - \int f(x)dx \qquad g(x)$ is "on top" on $\left[^-2, ^-1\right]$.

$$\text{Area } 1 = \frac{1}{3}x^3 - \left(\frac{1}{2}x^2 + 2x\right)\Big]_{-2}^{-1}$$

$$\text{Area } 1 = \left[\frac{1}{3}(^-1)^3 - \left(\frac{1}{2}(^-1)^2 + 2(^-1)\right)\right] - \left[\frac{1}{3}(^-2)^3 - \left(\frac{1}{2}(^-2)^2 + 2(^-2)\right)\right]$$

$$\text{Area } 1 = \left[\frac{^-1}{3} - \left(\frac{^-3}{2}\right)\right] - \left[\frac{^-8}{3} - (^-2)\right]$$

$$\text{Area } 1 = \left(\frac{7}{6}\right) - \left(\frac{^-2}{3}\right) = \frac{11}{6}$$

Area $2 = \int f(x)dx - \int g(x)dx \qquad f(x)$ is "on top" on $\left[^-1, 2\right]$.

$$\text{Area } 2 = \frac{1}{2}x^2 + 2x - \frac{1}{3}x^3\Big]_{-1}^{2}$$

$$\text{Area } 2 = \left(\frac{1}{2}(2)^2 + 2(2) - \frac{1}{3}(2)^3\right) - \left(\frac{1}{2}(^-1)^2 + 2(^-1) - \frac{1}{3}(^-1)^3\right)$$

$$\text{Area } 2 = \left(\frac{10}{3}\right) - \left(\frac{1}{2} - 2 + \frac{1}{3}\right)$$

$$\text{Area } 2 = \frac{27}{6}$$

Area $3 = \int g(x)dx - \int f(x)dx$ $g(x)$ is "on top" on $[2,3]$.

Area $3 = \frac{1}{3}x^3 - \left(\frac{1}{2}x^2 + 2x\right)\Big]_2^3$

Area $3 = \left[\frac{1}{3}(3)^3 - \left(\frac{1}{2}(3^2) + 2(3)\right)\right] - \left[\frac{1}{3}(2)^3 - \left(\frac{1}{2}(2)^2 + 2(2)\right)\right]$

Area $3 = \left(\frac{^-3}{2}\right) - \left(\frac{^-10}{3}\right) = \frac{11}{6}$

Total area $= \frac{11}{6} + \frac{27}{6} + \frac{11}{6} = \frac{49}{6} = 8\frac{1}{6}$

If you take the area bounded by a curve or curves and revolve it about a line, the result is a solid of revolution. To find the **volume** of such a solid, the Washer Method works in most instances. Imagine slicing through the solid perpendicular to the line of revolution. The "slice" should resemble a washer. Use an integral and the formula for the volume of disk.

$$Volume_{disk} = \pi \bullet radius^2 \bullet thickness$$

Depending on the situation, the radius is the distance from the line of revolution to the curve; or if there are two curves involved, the radius is the difference between the two functions. The thickness is *dx* if the line of revolution is parallel to the *x* axis and *dy* if the line of revolution is parallel to the *y* axis. Finally, integrate the volume expression using the boundary numbers from the interval.

Example:

Find the value of the solid of revolution found by revolving $f(x) = 9 - x^2$ about the *x* axis on the interval $[0,4]$.

$radius = 9 - x^2$
$thickness = dx$

$Volume = \int_0^4 \pi(9 - x^2)^2 dx$

Use the formula for volume of a disk.

$Volume = \pi \int_0^4 \left(81 - 18x^2 + x^4\right) dx$

$Volume = \pi \left(81x - \dfrac{18}{2+1}x^3 + \dfrac{1}{4+1}x^5\right)\Big]_0^4$

Take the integral.

$Volume = \pi \left(81x - 6x^3 + \dfrac{1}{5}x^5\right)\Big]_0^4$

Evaluate the integral first

$Volume = \pi \left[\left(324 - 384 + \dfrac{1024}{5}\right) - (0 - 0 + 0)\right]$

$x = 4$ then at $x = 0$

$Volume = \pi \left(144\dfrac{4}{5}\right) = 144\dfrac{4}{5}\pi = 454.9$

Solve problems using velocity and acceleration of a particle moving along a line.

If a particle (or a car, a bullet, etc.) is moving along a line, then the distance that the particle travels can be expressed by a function in terms of time.

1. The first derivative of the distance function will provide the velocity function for the particle. Substituting a value for time into this expression will provide the instantaneous velocity of the particle at the time. Velocity is the rate of change of the distance traveled by the particle.
   Taking the absolute value of the derivative provides the speed of the particle. A positive value for the velocity indicates that the particle is moving forward, and a negative value indicates the particle is moving backwards.

2. The second derivative of the distance function (which would also be the first derivative of the velocity function) provides the acceleration function. The acceleration of the particle is the rate of change of the velocity. If a value for time produces a positive acceleration, the particle is speeding up; if it produces a negative value, the particle is slowing down. If the acceleration is zero, the particle is moving at a constant speed. To find the time when a particle stops, set the first derivative (i.e. the velocity function) equal to zero and solve for time. This time value is also the instant when the particle changes direction.

Example:

The motion of a particle moving along a line is according to the equation: $s(t) = 20 + 3t - 5t^2$ where s is in meters and t is in time. Find the position, velocity, and acceleration of a particle at $t = 2$ seconds.

$s(2) = 20 + 3(2) - 5(2)^2$ Plug $t = 2$ into the original
    $= 6$ meters   equation to find the position.
$s'(t) = v(t) = 3 - 10t$ The derivative of the first function gives the velocity.

$v(2) = 3 - 10(2) = {}^-17$ m/s   Plug $t = 2$ into the velocity function to find the velocity. ${}^-17$ m/s indicates the particle is moving backwards.

$s''(t) = a(t) = {}^-10$ The second derivation of position gives
$a(2) = {}^-10$ m/s²   the acceleration. Substitute $t = 2$, yields an acceleration of ${}^-10$ m/s², which indicates the particle is slowing down.

Practice problem:

A particle moves along a line with acceleration $a(t) = 5t + 2$. The velocity after 2 seconds is ${}^-10$ m/sec.

1. Find the initial velocity.
2. Find the velocity at $t = 4$.

SUBAREA IV.   **GEOMETRY AND MEASUREMENT**

COMPETENCY 0009   UNDERSTAND THE PRINCIPLES OF MEASUREMENT.

Determine appropriate units and instruments for measuring a given quantity in a real-world context.

"When you can measure what you are speaking about and express it in numbers, you know something about it; but when you cannot measure it, when you cannot express it in numbers, your knowledge is of a meager and unsatisfactory kind." Lord Kelvin

Non-standard units are sometimes used when standard instruments might not be available. For example, students might measure the length of a room by their arm-spans. An inch originated as the length of three barley grains placed end to end. Seeds or stones might be used for measuring weight. In fact, our current "carat," used for measuring precious gems, was derived from carob seeds. In ancient times, baskets, jars and bowls were used to measure capacity.

To estimate measurement of familiar objects, it is first necessary to determine the units to be used.

Examples:
Length
1. The coastline of Florida   miles or kilometers
2. The width of a ribbon   inches or millimeters
3. The thickness of a book   inches or centimeters
4. The length of a football field   yards or meters
5. The depth of water in a pool   feet or meters

Weight or mass
1. A bag of sugar   pounds or grams
2. A school bus   tons or kilograms
3. A dime   ounces or grams

Capacity
1. Paint to paint a bedroom   gallons or liters
2. Glass of milk   cups or liters
3. Bottle of soda   quarts or liters
4. Medicine for child   ounces or milliliters

Estimate measurements, including length, area, volume, weight, time, temperature, and money.

It is necessary to be familiar with the metric and customary system in order to estimate measurements.

Some common equivalents include:

ITEM	APPROXIMATELY EQUAL TO	
	METRIC	IMPERIAL
large paper clip	1 gram	1 ounce
1 quart	1 liter	
average sized man	75 kilograms	170 pounds
1 yard	1 meter	
math textbook	1 kilogram	2 pounds
1 mile	1 kilometer	
1 foot	30 centimeters	
thickness of a dime	1 millimeter	0.1 inches

Estimate the measurement of the following items:

The length of an adult cow = _____ meters
The thickness of a compact disc = _____ millimeters
Your height = _____ meters
length of your nose = _____ centimeters
weight of your math textbook = _____ kilograms
weight of an automobile = _____ kilograms
weight of an aspirin = _____ grams

Make conversions within the metric or customary systems in a real-world context.

The units of **length** in the customary system are inches, feet, yards and miles.

> 12 inches (in.) = 1 foot (ft.)
> 36 in. = 1 yard (yd.)
> 3 ft. = 1 yd.
> 5280 ft. = 1 mile (mi.)
> 760 yd. = 1 mi.

To change from a **larger unit to a smaller unit, multiply.**
To change from a **smaller unit to a larger unit, divide.**

Example:
　　　4 mi. = _____ yd.
　　　Since 1760 yd. = 1 mile, multiply $4 \times 1760 = 7040$ yd.

Example:
　　　21 in. = _____ ft.
　　　$21 \div 12 = 1\frac{3}{4}$ ft.

The units of **weight** are ounces, pounds and tons.

> 16 ounces (oz.) = 1 pound (lb.)
> 　　　2,000 lb. = 1 ton (T.)

Example:　$2\frac{3}{4}$ T. = _____ lb.
　　　$2\frac{3}{4} \times 2,000 = 5,500$ lb.

The units of **capacity** are fluid ounces, cups, pints, quarts, and gallons.

8 fluid ounces (fl. oz.) = 1 cup (c.)
2 c. = 1 pint (pt.)
4 c. = 1 quart (qt.)
2 pt. = 1 qt.
4 qt. = 1 gallon (gal.)

larger → smaller
multiply

smaller → lgr
divide

Example1: 3 gal. = _____ qt.
$\qquad$ 3 × 4 = 12 qt.

Example: $1 \frac{1}{4}$ cups = _____ oz.
$\qquad$ $1\frac{1}{4}$ × 8 = 10 oz.

Example: 7 c. = _____ pt.
$\qquad$ 7 ÷ 2 = $3\frac{1}{2}$ pt.

Square units can be derived with knowledge of basic units of length by squaring the equivalent measurements.

1 square foot (sq. ft.) = 144 sq. in.
1 sq. yd. = 9 sq. ft.
1 sq. yd. = 1296 sq. in.

Example: 14 sq. yd. = _____ sq. ft.
$\qquad$ 14 × 9 = 126 sq. ft.

METRIC UNITS

The metric system is based on multiples of <u>ten</u>. Conversions are made by simply moving the decimal point to the left or right.

kilo-   1000 thousands
hecto-   100 hundreds
deca-   10 tens
nit
deci-   .1 tenths
centi-   .01 hundredths
milli-   .001 thousandths

The basic unit for **length** is the meter. One meter is approximately one yard.

The basic unit for **weight** or mass is the gram. A paper clip weighs about one gram.

The basic unit for **volume** is the liter. One liter is approximately a quart.

These are the most commonly used units.

1 m = 100 cm	1000 mL= 1 L	1000 mg = 1 g
1 m = 1000 mm	1 kL = 1000 L	1 kg = 1000 g
1 cm = 10 mm	1000 m = 1 km	

The prefixes are commonly listed from left to right for ease in conversion.

K H D U D C M

Example: 63 km = _____ m
Since there are 3 steps from <u>K</u>ilo to <u>U</u>nit, move the decimal point 3 places to the right.

63 km = 63,000 m

Example: 14 mL = _____ L
Since there are 3 steps from <u>M</u>illi to <u>U</u>nit, move the decimal point 3 places to the left.

14 mL = 0.014 L

Example: 56.4 cm = _____ mm
56.4 cm = 564 mm

Example: 9.1 m = _____ km
9.1 m = 0.0091 km

Example 5:   75 kg = _____ m
75 kg = 75,000,000 m

Apply the formulas for determining the circumferences and areas of circles in a real-world context.

The distance around a circle is the **circumference**. The ratio of the circumference to the diameter is represented by the Greek letter pi.

$\Pi \sim 3.14 \sim \dfrac{22}{7}$.

The circumference of a circle is found by the formula $C = 2\Pi r$ or $C = \Pi d$ where r is the radius of the circle and d is the diameter.

The **area** of a circle is found by the formula $A = \Pi r^2$.

<u>Example:</u> Find the circumference and area of a circle whose radius is 7 meters.

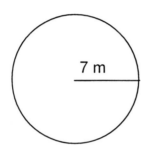

7 m

$C = 2\Pi r$ $A = \Pi r^2$
  $= 2(3.14)(7)$ $= 3.14(7)(7)$
  $= 43.96$ m $= 153.86$ m^2

Compute the area remaining when sections are cut   out of a given figure composed of triangles, squares, rectangles, parallelograms, trapezoids, or circles.

The strategy for solving problems of this nature should be to identify the given shapes and choose the correct formulas. Subtract the smaller cut out shape from the larger shape.

Sample problems:

1. Find the area of one side of the metal in the circular flat washer shown below:

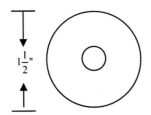

1. the shapes are both circles.

2. use the formula $A = \pi r^2$ for both.

    (Inside diameter is $3/8"$)

Area of larger circle   Area of smaller circle

$A = \pi r^2$ $A = \pi r^2$

$A = \pi(.75^2)$ $A = \pi(.1875^2)$

$A = 1.76625$ in^2 $A = .1104466$ in^2

Area of metal washer = larger area - smaller area

$$= 1.76625 \text{ in}^2 - .1104466 \text{ in}^2$$

$$= 1.6558034 \text{ in}^2$$

Find the perimeter or area of figures composed of parallelograms, triangles, circles, and trapezoids in a real-world context.

The **perimeter** of any polygon is the sum of the lengths of the sides.

P = sum of sides

Since the opposite sides of a rectangle are congruent, the perimeter of a rectangle equals twice the sum of the length and width or

P_{rect} = 2l + 2w or 2(l + w)

Similarly, since all the sides of a square have the same measure, the perimeter of a square equals four times the length of one side or

P_{square} = 4s

The **area** of a polygon is the number of square units covered by the figure.

A_{rect} = l × w
A_{square} = s^2

Example: Find the perimeter and the area of this rectangle.

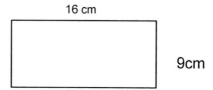

16 cm

9cm

P_{rect} = 2l + 2w
= 2(16) + 2(9)
= 32 + 18 = 50 cm

A_{rect} = l × w
= 16(9)
= 144 cm^2

Example: Find the perimeter and area of this square.

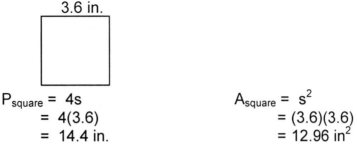

3.6 in.

$P_{square} = 4s$
$= 4(3.6)$
$= 14.4$ in.

$A_{square} = s^2$
$= (3.6)(3.6)$
$= 12.96$ in^2

In the following formulas, b = the base
and h = the height of an altitude drawn to the base.

$A_{parallelogram} = bh$
$A_{triangle} = \frac{1}{2}bh$
$A_{trapezoid} = \frac{1}{2}h(b_1 + b_2)$

Example: Find the area of a parallelogram whose base is 6.5 cm
and the height of the altitude to that base is 3.7 cm.

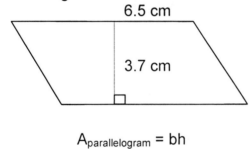

6.5 cm

3.7 cm

$A_{parallelogram} = bh$

$= (3.7)(6.5)$
$= 24.05$ cm^2

Example: Find the area of this triangle.

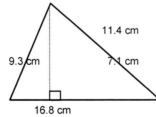

11.4 cm

9.3 cm   7.1 cm

16.8 cm

$A_{triangle} = \frac{1}{2}bh$

$= 0.5\,(16.8)\,(7.1)$
$= 59.64$ cm^2

Note that the altitude is drawn to the base measuring 16.8
cm. The lengths of the other two sides is unnecessary
information.

<u>Example:</u> Find the area of a right triangle whose sides measure 10 inches, 24 inches and 26 inches.

Since the hypotenuse of a right triangle must be the longest side, then the two perpendicular sides must measure 10 and 24 inches.

$$A_{triangle} = \tfrac{1}{2}bh$$
$$= \tfrac{1}{2}(10)(24)$$
$$= 120 \text{ sq. in.}$$

<u>Example:</u> Find the area of this trapezoid.

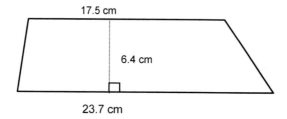

17.5 cm

6.4 cm

23.7 cm

The area of a trapezoid equals one-half the sum of the bases times the altitude.

$$A_{trapezoid} = \tfrac{1}{2}h(b_1 + b_2)$$
$$= 0.5(6.4)(17.5 + 23.7)$$
$$= 131.84 \text{ cm}^2$$

Compute the area remaining when sections are cut out of a given figure composed of triangles, squares, rectangles, parallelograms, trapezoids, or circles.

Example: You have decided to fertilize your lawn. The shapes and dimensions of your lot, house, pool and garden are given in the diagram below. The shaded area will not be fertilized. If each bag of fertilizer costs $7.95 and covers 4,500 square feet, find the total number of bags needed and the total cost of the fertilizer.

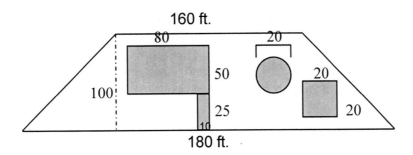

Area of Lot

$A = \frac{1}{2} h(b_1 + b_2)$

$A = \frac{1}{2}(100)(180 + 160)$

$A = 17,000$ sq ft

Area of House

$A = LW$

$A = (80)(50)$

$A = 4,000$ sq ft

Area of Driveway

$A = LW$

$A = (10)(25)$

$A = 250$ sq ft

Area of Pool

$A = \pi r^2$

$A = \pi (10)^2$

$A = 314.159$ sq. ft.

Area of Garden

$A = s^2$

$A = (20)^2$

$A = 400$ sq. ft.

Total area to fertilize = Lot area - (House + Driveway + Pool + Garden)

   = 17,000 - (4,000 + 250 + 314.159 + 400)

       = 12,035.841 sq ft

Number of bags needed = $\dfrac{\text{Total area to fertilize}}{4,500 \text{ sq.ft. bag}}$

   = $\dfrac{12,035.841}{4,500}$

   = 2.67 bags

Since we cannot purchase 2.67 bags we must purchase 3 full bags.

Total cost = Number of bags * $7.95

     = 3 * $7.95

     = $23.85

Apply the formulas for surface area and volume to right prisms, regular pyramids, right circular cylinders, cones, and spheres in a real-world context.

The **lateral** area is the area of the faces excluding the bases.

The **surface area** is the total area of all the faces, including the bases.

The **volume** is the number of cubic units in a solid. This is the amount of space a figure holds.

Right prism

$V = Bh$ (where B = area of the base of the prism and h = the height of the prism)

Rectangular right prism

$S = 2(lw + hw + lh)$ (where l = length, w = width and h = height)
$V = lwh$

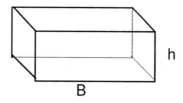

Example: Find the height of a box whose volume is 120 cubic meters and the area of the base is 30 square meters.

$$V = Bh$$
$$120 = 30h$$
$$h = 4 \text{ meters}$$

Regular pyramid

$V = 1/3Bh$

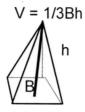

Right circular cylinder

$S = 2\Pi r(r + h)$ (where r is the radius of the base)
$V = \Pi r^2 h$

h

Right circular cone

$V = \frac{1}{3}Bh$

Solve real-world problems using money, rates, distance, time, temperature, and angle measures.

- Some problems can be solved using equations with rational expressions. First write the equation. To solve it, multiply each term by the LCD of all fractions. This will cancel out all of the denominators and give an equivalent algebraic equation that can be solved.

1. The denominator of a fraction is two less than three times the numerator. If 3 is added to both the numerator and denominator, the new fraction equals 1/2.

    original fraction: $\dfrac{x}{3x\text{-}2}$ revised fraction: $\dfrac{x+3}{3x+1}$

    $\dfrac{x+3}{3x+1} = \dfrac{1}{2}$ $2x + 6 = 3x + 1$

    $x = 5$

    original fraction: $\dfrac{5}{13}$

2. Elly Mae can feed the animals in 15 minutes. Jethro can feed them in 10 minutes. How long will it take them if they work together?

Solution: If Elly Mae can feed the animals in 15 minutes, then she could feed 1/15 of them in 1 minute, 2/15 of them in 2 minutes, $x/15$ of them in x minutes. In the same fashion Jethro could feed $x/10$ of them in x minutes. Together they complete 1 job. The equation is:

$$\frac{x}{15} + \frac{x}{10} = 1$$

Multiply each term by the LCD of 30:

$$2x + 3x = 30$$
$$x = 6 \text{ minutes}$$

3. A salesman drove 480 miles from Pittsburgh to Hartford. The next day he returned the same distance to Pittsburgh in half an hour less time than his original trip took, because he increased his average speed by 4 mph. Find his original speed.

Since distance = rate x time then time = $\dfrac{\text{distance}}{\text{rate}}$

original time – 1/2 hour = shorter return time

$$\frac{480}{x} - \frac{1}{2} = \frac{480}{x+4}$$

Multiplying by the LCD of $2x(x+4)$, the equation becomes:

$$480\left[2(x+4)\right]-1\left[x(x+4)\right]=480(2x)$$

$$960x+3840-x^2-4x=960x$$

$$x^2+4x-3840=0$$

$(x+64)(x-60)=0$ Either (x-60=0) or (x+64=0) or both=0

$x=60$ 60 mph is the original speed.

$x=64$ This is the solution since the time cannot be negative. Check your answer

$$\frac{480}{60}-\frac{1}{2}=\frac{480}{64}$$

$$8-\frac{1}{2}=7\frac{1}{2}$$

$$7\frac{1}{2}=7\frac{1}{2}$$

Try these:

1. Working together, Larry, Moe, and Curly can paint an elephant in 3 minutes. Working alone, it would take Larry 10 minutes or Moe 6 minutes to paint the elephant. How long would it take Curly to paint the elephant if he worked alone?

2. The denominator of a fraction is 5 more than twice the numerator. If the numerator is doubled, and the denominator is increased by 5, the new fraction is equal to 1/2. Find the original number.

3. A trip from Augusta, Maine to Galveston, Texas is 2108 miles. If one car drove 6 mph faster than a truck and got to Galveston 3 hours before the truck, find the speeds of the car and truck.

The union of all points on a simple closed surface and all points in its interior form a space figure called a **solid**. The five regular solids, or **polyhedra**, are the cube, tetrahedron, octahedron, icosahedron, and dodecahedron. A **net** is a two-dimensional figure that can be cut out and folded up to make a three-dimensional solid. Below are models of the five regular solids with their corresponding face polygons and nets.

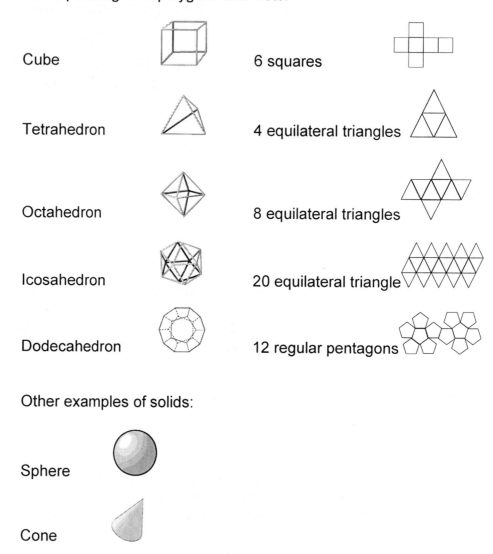

Cube   6 squares

Tetrahedron   4 equilateral triangles

Octahedron   8 equilateral triangles

Icosahedron   20 equilateral triangle

Dodecahedron   12 regular pentagons

Other examples of solids:

Sphere

Cone

COMPETENCY 0010   UNDERSTAND PRINCIPLES OF EUCLIDEAN GEOMETRY.

Writing the inverse, converse, and contrapositive of a given conditional.

Conditional statements are explained in section 30.1.

Conditional: If p, then q

> p is the hypothesis. q is the conclusion.

Inverse: If ~ p, then ~ q

> Negate both the hypothesis (If not p, then not q) and the conclusion from the original conditional.

Converse : If q, then p.

> Reverse the 2 clauses. The original hypothesis becomes the conclusion. The original conclusion then becomes the new hypothesis.

Contrapositive: If ~ q, then ~ p.

> Reverse the 2 clauses. The If not q, then not p original hypothesis becomes the conclusion. The original conclusion then becomes the new hypothesis. THEN negate both the new hypothesis and the new conclusion.

<u>Example</u>: Given the **conditional**:

If an angle has 60°, then it is an acute angle.

Its **inverse**, in the form "If ~ p, then ~ q", would be:

If an angle doesn't have 60°, then it is not an acute angle.

> NOTICE that the inverse is not true, even though the conditional statement was true.

Its **converse**, in the form "If q, then p", would be:

If an angle is an acute angle, then it has 60°.

> NOTICE that the converse is not true, even though the conditional statement was true.

Its **contrapositive**, in the form "If ~q, then ~p", would be:

If an angle isn't an acute angle, then it doesn't have 60°.

NOTICE that the contrapositive is true, assuming the original conditional statement was true.

TIP: If you are asked to pick a statement that is logically equivalent to a given conditional, look for the contra-positive. The inverse and converse are not always logically equivalent to every conditional. The contra-positive is ALWAYS logically equivalent.

Find the inverse, converse and contrapositive of the following conditional statement. Also determine if each of the 4 statements is true or false.

Conditional: If $x = 5$, then $x^2 - 25 = 0$.　　　TRUE
Inverse: If $x \neq 5$, then $x^2 - 25 \neq 0$.　　　FALSE, x could be ⁻5
Converse: If $x^2 - 25 = 0$, then $x = 5$.　　　FALSE, x could be ⁻5
Contrapositive: If $x^2 - 25 \neq 0$, then $x \neq 5$.　　　TRUE
Conditional: If $x = 5$, then $6x = 30$.　　　TRUE
Inverse:　If $x \neq 5$, then $6x \neq 30$.　　　TRUE
Converse:　If $6x = 30$, then $x = 5$.　　　TRUE
Contrapositive: If $6x \neq 30$, then $x \neq 5$.　　　TRUE

Sometimes, as in this example, all 4 statements can be logically equivalent; however, the only statement that will always be logically equivalent to the original conditional is the contrapositive.

Identify point, line, and plane as undefined terms and use symbols for lines, segments, rays, and distances.

The 3 undefined terms of geometry are point, line, and plane.

A plane is a flat surface that extends forever in two dimensions. It has no ends or edges. It has no thickness to it. It is usually drawn as a parallelogram that can be named either by 3 non-collinear points (3 points that are not on the same line) on the plane or by placing a letter in the corner of the plane that is not used elsewhere in the diagram.

A line extends forever in one dimension. It is determined and named by 2 points that are on the line. The line consists of every point that is between those 2 points as well as the points that are on the "straight" extension each way. A line is drawn as a line segment with arrows facing opposite directions on each end to indicate that the line continues in both directions forever.

A point is a position in space, on a line, or on a plane. It has no thickness and no width. Only 1 line can go through any 2 points. A point is represented by a dot named by a single letter.

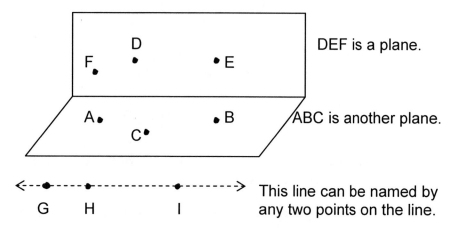

DEF is a plane.

ABC is another plane.

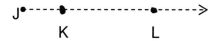

G   H   I

This line can be named by any two points on the line.

It could be named $\overleftrightarrow{GH}$, $\overleftrightarrow{HI}$, $\overleftrightarrow{GI}$, $\overleftrightarrow{IG}$, $\overleftrightarrow{IH}$, or $\overleftrightarrow{HG}$. Any 2 points (letters) on the line can be used and their order is not important in naming a line.

In the above diagrams, A, B, C, D, E, F, G, H, and I are all locations of individual points.

A ray is not an undefined term. A ray consists of all the points on a line starting at one given point and extending in only one of the two opposite directions along the line. The ray is named by naming 2 points on the ray. The first point must be the endpoint of the ray, while the second point can be any other point along the ray. The symbol for a ray is a ray above the 2 letters used to name it. The endpoint of the ray MUST be the first letter.

J•- - - - -•- - - - - - - - - - - -•- - - - - ->
          K   L

This ray could be named $\overrightarrow{JK}$ or $\overrightarrow{JL}$. It can not be called $\overrightarrow{KJ}$ or $\overrightarrow{LJ}$ or $\overrightarrow{LK}$ or $\overrightarrow{KL}$ because none of those names start with the endpoint, J.

The **distance** between 2 points on a number line is equal to the absolute value of the difference of the two numbers associated with the points.

If one point is located at "*a*" and the other point is at "*b*", then the distance between them is found by this formula:

$$\text{distance} = |a - b| \text{ or } |b - a|$$

If one point is located at $^-3$ and another point is located at 5, the distance between them is found by:

$$\text{distance} = |a - b| = |(^-3) - 5| = |^-8| = 8$$

Apply the inequality relationships among the angles and sides of a triangle.

The Triangle Inequality Theorem shows that the sum of the length of any two sides is greater than the length of the remaining side. So that, in the triangle below,

$a + b > c$
$a + c > b$
$b + c > a$.

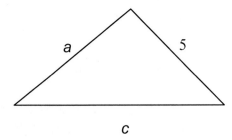

If a triangle had an unknown side, *p*, the Triangle Inequality Theorem would be applied to determine a reasonable range of possible values for the unknown side.

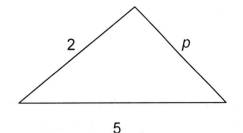

$2 + p > 5$
$2 + 5 > p$
$p + 5 > 2$

The expressions could be arranged to show: $p > 5\text{-}2$ or $p > 3$
$7 > p$
$p > \text{-}3$.

The final expression is not valid since a side must not be a negative number. The other two expressions show that side *p* is a value between 3 and 5.

The side of the triangle that is opposite the largest angle is the longer side.   We can use this rule to determine a reasonable measurement of the third unknown angle.

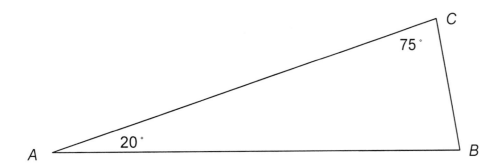

Knowing that the sum of all angles in a triangle is equal to 180˚, we can determine that Angle C is equal to 85˚.

Angle A + Angle B + Angle C = 180˚
20˚+Angle B + 75˚= 180˚
Angle B = 180˚-20˚-75˚
Angle B = 85˚

Given the known angles, Angle A < Angle C < Angle B

To order the sides according to size, we can apply the angle-side relationship.

$\overline{BC} < \overline{AB} < \overline{AC}$

Use the SAS, ASA, and SSS postulates to show pairs of triangles congruent, including the case of overlapping triangles.

Two triangles can be proven congruent by comparing pairs of appropriate congruent corresponding parts.

SSS POSTULATE

If three sides of one triangle are congruent to three sides of another triangle, then the two triangles are congruent.

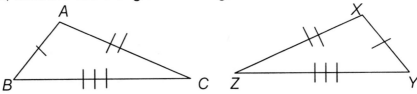

Since $AB \cong XY$, $BC \cong YZ$ and $AC \cong XZ$, then $\triangle ABC \cong \triangle XYZ$.

Example: Given isosceles triangle *ABC* with *D* the midpoint of base *AC*, prove the two triangles formed by *AD* are congruent.

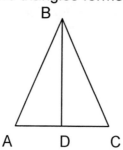

Proof:
1. Isosceles triangle *ABC*,
   D midpoint of base *AC*   Given
2. $AB \cong BC$ An isosceles △ has two congruent sides
3. $AD \cong DC$ Midpoint divides a line into two equal parts
4. $BD \cong BD$ Reflexive
5. △ *ABD* ≅ △*BCD*   SSS

SAS POSTULATE

If two sides and the included angle of one triangle are congruent to two sides and the included angle of another triangle, then the two triangles are congruent.

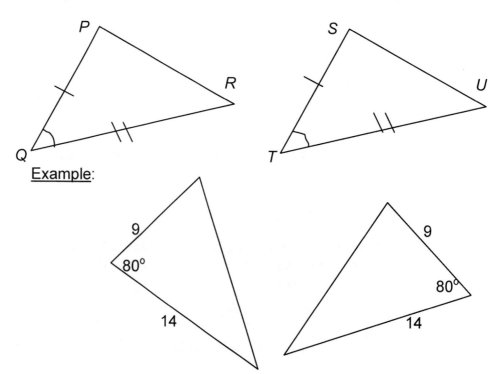

Example:

The two triangles are congruent by SAS.

ASA POSTULATE

If two angles and the included side of one triangle are congruent to two angles and the included side of another triangle, the triangles are congruent.

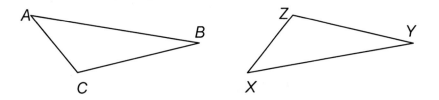

$\angle A \cong \angle X$, $\angle B \cong \angle Y$, $AB \cong XY$ then $\triangle ABC \cong \triangle XYZ$ by ASA

Example: Given two right triangles with one leg of each measuring 6 cm and the adjacent angle 37°, prove the triangles are congruent.

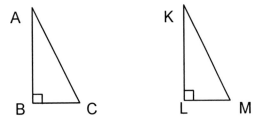

1. Right triangles *ABC* and *KLM*   Given
   $AB = KL = 6$ cm
   $\angle A = \angle K = 37°$
2. $AB \cong KL$ Figures with the same
   $\angle A \cong \angle K$ measure are congruent
3. $\angle B \cong \angle L$ All right angles are
                                                       congruent.
4. $\triangle ABC \cong \triangle KLM$ ASA

Example:
What method would you use to prove the triangles congruent?

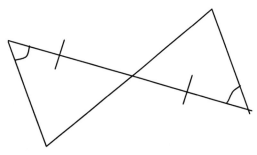

ASA because vertical angles are congruent.

Solve real-world problems involving similar or congruent figures.

Example:

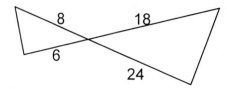

The two triangles are similar since the sides are proportional and vertical angles are congruent.

Example: Given two similar quadrilaterals. Find the lengths of sides *x, y,* and *z.*

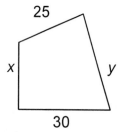
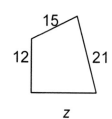

Since corresponding sides are proportional:

$$\frac{15}{25} = \frac{3}{5} \text{ so the scale is } \frac{3}{5}$$

$$\frac{12}{x} = \frac{3}{5} \qquad\qquad \frac{21}{y} = \frac{3}{5} \qquad\qquad \frac{z}{30} = \frac{3}{5}$$

$$3x = 60 \qquad\qquad 3y = 105 \qquad\qquad 5z = 90$$
$$x = 20 \qquad\qquad y = 35 \qquad\qquad z = 18$$

Polygons are similar if and only if there is a one-to-one correspondence between their vertices such that the corresponding angles are congruent and the lengths of corresponding sides are proportional.

Given the rectangles below, compare the area and perimeter.

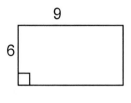

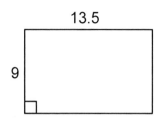

$A = LW$	$A = LW$	1. write formula
$A = (6)(9)$	$A = (9)(13.5)$	2. substitute known values
$A = 54$ sq. units	$A = 121.5$ sq. units	3. compute

$P = 2(L + W)$	$P = 2(L + W)$	1. write formula
$P = 2(6 + 9)$	$P = 2(9 + 13.5)$	2. substitute known values
$P = 30$ units	$P = 45$ units	3. compute

Notice that the areas relate to each other in the following manner:

Ratio of sides $9/13.5 = 2/3$

Multiply the first area by the square of the reciprocal $(3/2)^2$ to get the second area.

$$54 \times (3/2)^2 = 121.5$$

The perimeters relate to each other in the following manner:

Ratio of sides $9/13.5 = 2/3$

Multiply the perimeter of the first by the reciprocal of the ratio to get the perimeter of the second.

$$30 \times 3/2 = 45$$

Solve real-world problems applying the Pythagorean theorem and its converse.

Pythagorean theorem states that the square of the length of the hypotenuse is equal to the sum of the squares of the lengths of the legs. Symbolically, this is stated as:

$$c^2 = a^2 + b^2$$

Given the right triangle below, find the missing side.

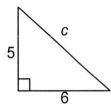

$c^2 = a^2 + b^2$ 1. write formula

$c^2 = 5^2 + 6^2$ 2. substitute known values

$c^2 = 61$ 3. take square root

$c = \sqrt{61}$ or 7.81   4. solve

The Converse of the Pythagorean Theorem states that if the square of one side of a triangle is equal to the sum of the squares of the other two sides, then the triangle is a right triangle.

Example:
Given ΔXYZ, with sides measuring 12, 16 and 20 cm. Is this a right triangle?

$$c^2 = a^2 + b^2$$
$$20^2 \ \underline{?} \ 12^2 + 16^2$$
$$400 \ \underline{?} \ 144 + 256$$
$$400 \ = 400$$

Yes, the triangle is a right triangle.

This theorem can be expanded to determine if triangles are obtuse or acute.

If the square of the longest side of a triangle is greater than the sum of the squares of the other two sides, then the triangle is an obtuse triangle.
and
If the square of the longest side of a triangle is less than the sum of the squares of the other two sides, then the triangle is an acute triangle.

Calculate the length of an arc and the area of sectors of circle.

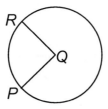

$$\frac{\measuredangle PQR}{360°} = \frac{\text{length of arc } RP}{\text{circumference of } \bigcirc Q} = \frac{\text{area of sector } PQR}{\text{area of } \bigcirc Q}$$

While an arc has a measure associated to the degree measure of a central angle, it also has a length which is a fraction of the circumference of the circle.

For each central angle and its associated arc, there is a sector of the circle which resembles a pie piece. The area of such a sector is a fraction of the area of the circle.

The fractions used for the area of a sector and length of its associated arc are both equal to the ratio of the central angle to 360°.

Examples:

1.

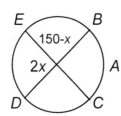

$\odot A$ has a radius of 4 cm. What is the length of arc *ED*?

$$2x + 150 - x = 180$$
$$x + 150 = 180$$
$$x = 30$$

Arc *BE* and arc *DE* make a semicircle.

$$\text{Arc } ED = 2(30) = 60°$$

The ratio 60° to 360° is equal to the ratio of arch length *ED* to the circumference of $\odot A$.

$$\frac{60}{360} = \frac{\text{arc length } ED}{2\pi 4}$$

$$\frac{1}{6} = \frac{\text{arc length}}{8\pi}$$

Cross multiply and solve for the arc length.

$$\frac{8\pi}{6} = \text{arc length}$$

$$\text{arc length } ED = \frac{4\pi}{3} \text{ cm}.$$

2.

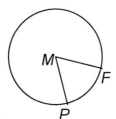

The radius of ○ M is 3 cm. The length of arc *PF* is 2π cm. What is the area of sector *PMF*?

Find the circumference and area of the circle.

Circumference of ○ $M = 2\pi(3) = 6\pi$

Area of ○$M = \pi(3)^2 = 9\pi$

$$\frac{\text{area of } PMF}{9\pi} = \frac{2\pi}{6\pi}$$

The ratio of the sector area to the circle area is the same as the arc length to the circumference.

$$\frac{\text{area of } PMF}{9\pi} = \frac{1}{3}$$

$$\text{area of } PMF = \frac{9\pi}{3}$$

$$\text{area of } PMF = 3\pi$$

Solve for the area of the sector.

Apply the theorems pertaining to the relationships of chords, diameters, radii, and tangents with respect to circles and to each other.

A tangent line intersects a circle in exactly one point. If a radius is drawn to that point, the radius will be perpendicular to the tangent.

A chord is a segment with endpoints on the circle. If a radius or diameter is perpendicular to a chord, the radius will cut the chord into two equal parts.

If two chords in the same circle have the same length, the two chords will have arcs that are the same length, and the two chords will be equidistant from the center of the circle. Distance from the center to a chord is measured by finding the length of a segment from the center perpendicular to the chord.

Examples:

1.

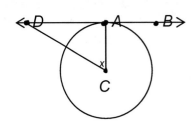

$\overline{DB}$ is tangent to $\bigcirc C$ at A.

$m\angle ADC = 40°$. Find x

$\overline{AC} \perp \overline{DB}$ A radius is $\perp$ to a tangent at the point of tangency.

$m\angle DAC = 90°$ Two segments that are $\perp$ form a $90°$ angle.

$40 + 90 + x = 180$ The sum of the angles of a triangle is $180°$.

$x = 50°$ Solve for x.

2.

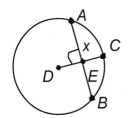

$\overline{CD}$ is a radius and $\overline{CD} \perp$ chord $\overline{AB}$.

$\overline{AB} = 10$. Find x.

$x = \dfrac{1}{2}(10)$

$x = 5$

If a radius is $\perp$ to a chord, the radius bisects the chord.

Angles related to circles.

Angles with their vertices on the circle:

An inscribed angle is an angle whose vertex is on the circle. Such an angle could be formed by two chords, two diameters, two secants, or a secant and a tangent. An inscribed angle has one arc of the circle in its interior. The measure of the inscribed angle is one-half the measure of this intercepted arc. If two inscribed angles intercept the same arc, the two angles are congruent (i.e. their measures are equal). If an inscribed angle intercepts an entire semicircle, the angle is a right angle.

Angles with their vertices in a circle's interior:

When two chords intersect inside a circle, two sets of vertical angles are formed. Each set of vertical angles intercepts two arcs which are across from each other. The measure of an angle formed by two chords in a circle is equal to one-half the sum of the angle intercepted by the angle and the arc intercepted by its vertical angle.

Angles with their vertices in a circle's exterior:

If an angle has its vertex outside of the circle and each side of the circle intersects the circle, then the angle contains two different arcs. The measure of the angle is equal to one-half the difference of the two arcs.

Examples:

1.

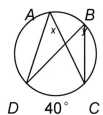

Find x and y.
arc $DC = 40°$

$$m\angle DAC = \frac{1}{2}(40) = 20°$$ $\angle DAC$ and $\angle DBC$ are both

$$m\angle DBC = \frac{1}{2}(40) = 20°$$ inscribed angles, so each one

$$x = 20° \text{ and } y = 20°$$ has a measure equal to one-half the measure of arc DC.

Lengths of chords, secants, and tangents

Intersecting chords:

If two chords intersect inside a circle, each chord is divided into two smaller segments. The product of the lengths of the two segments formed from one chord equals the product of the lengths of the two segments formed from the other chord.

Intersecting tangent segments:

If two tangent segments intersect outside of a circle, the two segments have the same length.

Intersecting secant segments:

If two secant segments intersect outside a circle, a portion of each segment will lie inside the circle and a portion (called the exterior segment) will lie outside the circle. The product of the length of one secant segment and the length of its exterior segment equals the product of the length of the other secant segment and the length of its exterior segment.

Tangent segments intersecting secant segments:

If a tangent segment and a secant segment intersect outside a circle, the square of the length of the tangent segment equals the product of the length of the secant segment and its exterior segment.

Examples:

1.

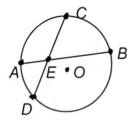

$\overline{AB}$ and $\overline{CD}$ are chords.
CE=10, ED=x, AE=5, EB=4

$(AE)(EB) = (CE)(ED)$ Since the chords intersect in the circle, the products of the segment pieces are equal.

$5(4) = 10x$
$20 = 10x$
$x = 2$ Solve for x.

2.

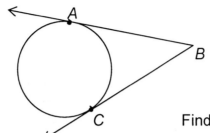

$\overline{AB}$ and $\overline{CD}$ are chords.
$\overline{AB} = x^2 + x - 2$
$\overline{BC} = x^2 - 3x + 5$
Find the length of $\overline{AB}$ and $\overline{BC}$.

$\overline{AB} = x^2 + x - 2$ $\overline{BC} = x^2 - 3x + 5$	Given
$\overline{AB} = \overline{BC}$	Intersecting tangents are equal.
$x^2 + x - 2 = x^2 - 3x + 5$	Set the expression equal to each other and solve.
$4x = 7$	Substitute and solve.
$x = 1.75$	

$(1.75)^2 + 1.75 - 2 = \overline{AB}$

$\overline{AB} = \overline{BC} = 2.81$

Use appropriate problem solving strategies to find the solution.

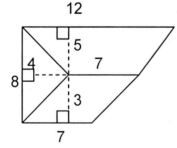

1. Find the area of the given figure.

2. Cut the figure into familiar shapes.

3. Identify what type figures are given and write the appropriate formulas.

Area of figure 1 (triangle)	Area of figure 2 (parallelogram)	Area of figure 3 (trapezoid)
$A = \frac{1}{2}bh$	$A = bh$	$A = \frac{1}{2}h(a+b)$
$A = \frac{1}{2}(8)(4)$	$A = (7)(3)$	$A = \frac{1}{2}(5)(12+7)$
$A = 16$ sq. ft	$A = 21$ sq. ft	$A = 47.5$ sq. ft

Now find the total area by adding the area of all figures.

Total area $= 16 + 21 + 47.5$
Total area $= 84.5$ square ft

Identify and use the parts of a regular polygon to compute the area of a polygon.

Given the figure below, find the area by dividing the polygon into smaller shapes.

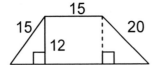

1. divide the figure into two triangles and a rectangle.

2. find the missing lengths.

3. find the area of each part.

4. find the sum of all areas.

Find base of both right triangles using Pythagorean Formula:

$$a^2 + b^2 = c^2$$
$$a^2 + 12^2 = 15^2$$
$$a^2 = 225 - 144$$
$$a^2 = 81$$
$$a = 9$$

$$a^2 + b^2 = c^2$$
$$a^2 + 12^2 = 20^2$$
$$a^2 = 400 - 144$$
$$a^2 = 256$$
$$a = 16$$

Area of triangle 1   Area of triangle 2   Area of rectangle

$$A = \frac{1}{2}bh$$
$$A = \frac{1}{2}(9)(12)$$
$$A = 54 \text{ sq. units}$$

$$A = \frac{1}{2}bh$$
$$A = \frac{1}{2}(16)(12)$$
$$A = 96 \text{ sq. units}$$

$$A = LW$$
$$A = (15)(12)$$
$$A = 180 \text{ sq. units}$$

Find the sum of all three figures.

$$54 + 96 + 180 = 330 \text{ square units}$$

Compare the perimeters and areas of similar polygons.

Polygons are similar if and only if there is a one-to-one correspondence between their vertices such that the corresponding angles are congruent and the lengths of corresponding sides are proportional.

Given the rectangles below, compare the area and perimeter.

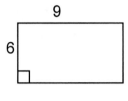   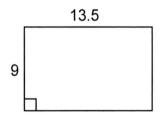

$A = LW$	$A = LW$	1. write formula
$A = (6)(9)$	$A = (9)(13.5)$	2. substitute known values
$A = 54$ sq. units	$A = 121.5$ sq. units	3. compute
$P = 2(L + W)$	$P = 2(L + W)$	1. write formula
$P = 2(6 + 9)$	$P = 2(9 + 13.5)$	2. substitute known values
$P = 30$ units	$P = 45$ units	3. compute

Notice that the areas relate to each other in the following manner:

   Ratio of sides $9/13.5 = 2/3$

Multiply the first area by the square of the reciprocal $(3/2)^2$ to get the second area.
$$54 \times (3/2)^2 = 121.5$$

   The perimeters relate to each other in the following manner:

   Ratio of sides $9/13.5 = 2/3$

Multiply the perimeter of the first by the reciprocal of the ratio to get the perimeter of the second.

$$30 \times 3/2 = 45$$

Apply the formulas for lateral area, total area and volume to right prisms and regular pyramids.

FIGURE	LATERAL AREA	TOTAL AREA	VOLUME
Right prism	sum of area of lateral faces (rectangles)	lateral area plus 2 times the area of base	area of base times height
regular pyramid	sum of area of lateral faces (triangles)	lateral area plus area of base	1/3 times the area of the base times the height

Find the total area of the given figure:

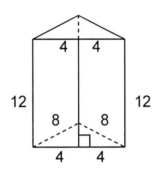

1. Since this is a triangular prism, first find the area of the bases.

2. Find the area of each rectangular lateral face.

3. Add the areas together.

$A = \frac{1}{2}bh$ $A = LW$ 1. write formula

$8^2 = 4^2 + h^2$ 2. find the height of
$h = 6.928$ the base triangle

$A = \frac{1}{2}(8)(6.928)$ $A = (8)(12)$

                                        3. substitute known
                                            values
$A = 27.713$ sq. units $A = 96$ sq. units   4. compute

Total Area $= 2(27.713) + 3(96)$
          $= 343.426$ sq. units

Apply the formulas for lateral area, total area and volume to right circular cylinders and cones.

FIGURE	VOLUME	TOTAL SURFACE AREA	LATERAL AREA
Right Cylinder	$\pi r^2 h$	$2\pi rh + 2\pi r^2$	$2\pi rh$
Right Cone	$\dfrac{\pi r^2 h}{3}$	$\pi r\sqrt{r^2 + h^2} + \pi r^2$	$\pi r\sqrt{r^2 + h^2}$

Note: $\sqrt{r^2 + h^2}$ is equal to the slant height of the cone.

Sample problem:

1. A water company is trying to decide whether to use traditional cylindrical paper cups or to offer conical paper cups since both cost the same. The traditional cups are 8 cm wide and 14 cm high. The conical cups are 12 cm wide and 19 cm high. The company will use the cup that holds the most water.

1. Draw and label a sketch of each.

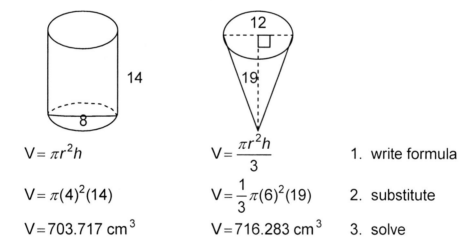

$V = \pi r^2 h$	$V = \dfrac{\pi r^2 h}{3}$	1. write formula
$V = \pi (4)^2 (14)$	$V = \dfrac{1}{3}\pi (6)^2 (19)$	2. substitute
$V = 703.717 \text{ cm}^3$	$V = 716.283 \text{ cm}^3$	3. solve

The choice should be the conical cup since its volume is more.

Apply the formulas for surface area and volume of spheres.

FIGURE	VOLUME	TOTAL SURFACE AREA
Sphere	$\dfrac{4}{3}\pi r^3$	$4\pi r^2$

Sample problem:

1. How much material is needed to make a basketball that has a diameter of 15 inches? How much air is needed to fill the basketball?

Draw and label a sketch:

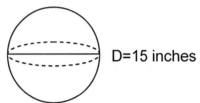 D=15 inches

Total surface area   Volume

$TSA = 4\pi r^2$ $V = \dfrac{4}{3}\pi r^3$ 1. write formula

$= 4\pi(7.5)^2$ $= \dfrac{4}{3}\pi(7.5)^3$ 2. substitute

$= 706.858 \text{ in}^2$ $= 1767.1459 \text{ in}^3$ 3. solve

COMPETENCY 0011   UNDERSTAND COORDINATE AND
TRANSFORMATIONAL GEOMETRY.

Identify transformations, dilations, or symmetry of geometric figures.

A **transformation** is a change in the position, shape, or size of a geometric figure. **Transformational geometry** is the study of manipulating objects by flipping, twisting, turning and scaling. **Symmetry** is exact similarity between two parts or halves, as if one were a mirror image of the other.

A **translation** is a transformation that "slides" an object a fixed distance in a given direction. The original object and its translation have the same shape and size, and they face in the same direction.

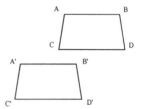

An example of a translation in architecture would be stadium seating. The seats are the same size and the same shape and face in the same direction.

A **rotation** is a transformation that turns a figure about a fixed point called the center of rotation. An object and its rotation are the same shape and size, but the figures may be turned in different directions. Rotations can occur in either a clockwise or a counterclockwise direction.

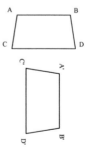

Rotations can be seen in wallpaper and art, and a Ferris wheel is an example of rotation.

An object and its **reflection** have the same shape and size, but the figures face in opposite directions.

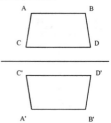

The line (where a mirror may be placed) is called the **line of reflection**. The distance from a point to the line of reflection is the same as the distance from the point's image to the line of reflection.

A **glide reflection** is a combination of a reflection and a translation.

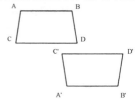

Another type of transformation is **dilation**. Dilation is a transformation that "shrinks" or "makes it bigger."

Example:
Using dilation to transform a diagram.

Starting with a triangle whose center of dilation is point P,

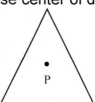

we dilate the lengths of the sides by the same factor to create a new triangle.

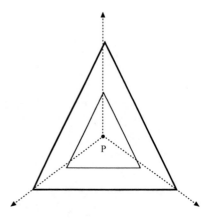

Determine the center of and the radius and graph a circle, given its equation.

The equation of a circle with its center at (h, k) and a radius r units is:

$$(x - h)^2 + (y - k)^2 = r^2$$

Sample Problem:

1. Given the equation $x^2 + y^2 = 9$, find the center and the radius of the circle. Then graph the equation.

First, writing the equation in standard circle form gives:

$$(x - 0)^2 + (y - 0)^2 = 3^2$$

therefore, the center is (0,0) and the radius is 3 units.

Sketch the circle:

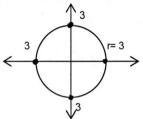

2. Given the equation $x^2 + y^2 - 3x + 8y - 20 = 0$, find the center and the radius. Then graph the circle.

First, write the equation in standard circle form by completing the square for both variables.

$x^2 + y^2 - 3x + 8y - 20 = 0$ 1. Complete the squares.

$(x^2 - 3x + 9/4) + (y^2 + 8y + 16) = 20 + 9/4 + 16$

$(x - 3/2)^2 + (y + 4)^2 = 153/4$

The center is $(3/2, {}^-4)$ and the radius is $\dfrac{\sqrt{153}}{2}$ or $\dfrac{3\sqrt{17}}{2}$.

Graph the circle.

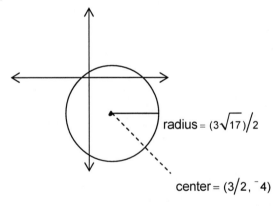

radius $= \left(3\sqrt{17}\right)/2$

center $= \left(3/2, {}^-4\right)$

Identify the equation of a circle, given its center and radius.

To write the equation given the center and the radius use the standard form of the equation of a circle:

$$(x - h)^2 + (y - k)^2 = r^2$$

Sample problems:

Given the center and radius, write the equation of the circle.

1. Center ($^-$1,4); radius 11

$(x - h)^2 + (y - k)^2 = r^2$	1. Write standard equation.
$(x - (^-1))^2 + (y - (4))^2 = 11^2$	2. Substitute.
$(x + 1)^2 + (y - 4)^2 = 121$	3. Simplify.

2. Center ($\sqrt{3}, ^-1/2$); radius $= 5\sqrt{2}$

$(x - h)^2 + (y - k)^2 = r^2$	1. Write standard equation.
$(x - \sqrt{3})^2 + (y - (^-1/2))^2 = (5\sqrt{2})^2$	2. Substitute.
$(x - \sqrt{3})^2 + (y + 1/2)^2 = 50$	3. Simplify.

Conic sections result from the intersection of a cone and a plane. The three main types of conics are parabolas, ellipses, and hyperbolas.

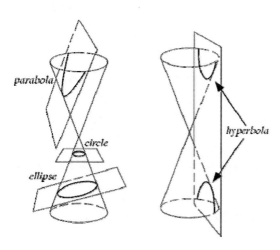

The general equation for a conic section is:

$$Ax^2 + Bxy + Cy^2 + Dx + Ey + F = 0$$

The value of $B^2 - 4AC$ determines the type of conic. If $B^2 - 4AC$ is less than zero the curve is an ellipse or a circle. If equal to zero, the curve is a parabola. If greater than zero, the curve is a hyperbola.

A **parabola** is the set of all points in a plane equidistant from a fixed point called the focus (F) and a fixed line called the directrix. The vertex is the point halfway between the focus and the directrix that lies on the parabola.

The equation of a parabola with focus $(h, k + p)$, directrix $y = k - p$, and vertex (h, k) is

$$(x - h)^2 = 4p(y - k)$$

If $p > 0$ the parabola opens up. If $p < 0$ the parabola opens down.

Example:

Find the equation of a parabola with vertex (1,2) and focus (1,3).

$k + p = 2$	Definition of focus.
$1 + p = 2$	Substitute for k (vertex value).
$p = 1$	Solve for p.
$(x - 1)^2 = 4(p)(y - 2)$	Equation of parabola
$(x - 1)^2 = 4(y - 2)$	Substitute for p.

An **ellipse** is the set of all points in a plane the sum of whose distances from two fixed points called foci is equal. The vertices of an ellipse are the two points farthest from the center. The major axis joins the vertices. The general equation of an ellipse with center located at (h, k), vertices $(h \pm a, k)$, foci $(h \pm c, k)$

$$\frac{(x - h)^2}{a^2} + \frac{(y - k)^2}{b^2} = 1 \quad \text{where } a \geq b > 0 \text{ and } a^2 - b^2 = c^2$$

The distance between the foci is 2c. The length of the major axis is 2a.

Example:

Find the equation of an ellipse with foci (4, -4), (6, -4) and vertices (3, -2), (7, -2).

The length of the major axis joining the vertices (3, -2), (7, -2) is 4, so a = 2. The distance between the foci (4, -4), (6, -4) is 2, so c = 1. Therefore, $b^2 = a^2 - c^2 = 3$. The center of the ellipse is (5, -2).

Substituting the resolved values into the equation of an ellipse yields...

$$\frac{(x-5)^2}{4} + \frac{(y+2)^2}{3} = 1$$

A **hyperbola** is the set of all points in a plane the difference of whose distances from two fixed points called foci is equal. The vertices of a hyperbola are the two points where the curve makes its sharpest turns located on the major axis (the line through the foci). The general equation of a hyperbola centered at (h, k), foci (h$\pm$c, k), vertices (h$\pm$a, k) and asymptotes $y - k = \pm\frac{b}{a}(x - h)$

$$\frac{(x-h)^2}{a^2} - \frac{(y-k)^2}{b^2} = 1 \quad \text{where } a^2 + b^2 = c^2$$

The distance between the foci is 2c. The distance between the vertices is 2a.

Example:

Find the equation of a hyperbola with foci (1, 3) and (7, 3) and vertices (2, 3) and (6,3).

The distance between the foci is 6, thus c = 3. The distance between the vertices is 4, thus a = 2. Therefore, $b^2 = c^2 - a^2 = 3^2 - 2^2 = 5$. The hyperbola is centered at (4, 3).

Substituting the resolved values into the equation of a hyperbola yields...

$$\frac{(x-4)^2}{4} - \frac{(y-3)^2}{5} = 1$$

The following table illustrates the properties of each quadrilateral.

	Parallel Opposite Sides	Bisecting Diagonals	Equal Opposite Sides	Equal Opposite Angles	Equal Diagonals	All Sides Equal	All Angles Equal	Perpendicular Diagonals
Parallelogram	X	X	X	X				
Rectangle	X	X	X	X	X		X	
Rhombus	X	X	X	X		X		X
Square	X	X	X	X	X	X	X	X

PARALLELOGRAMS

A parallelogram is a quadrilateral with both pairs of opposite sides parallel.

Theorems:

1. Consecutive pairs of angles are supplementary.
2. Opposite angles are congruent.
3. A diagonal divides the parallelogram into two congruent triangles.
4. Opposite sides are congruent
5. Diagonals bisect each other.

<u>Example:</u>

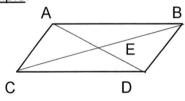

Given: Parallelogram ABCD, with diagonals AD and BC intersecting at E
Prove: AE ≅ DE

1. Parallelogram ABCD, with diagonals AC and BD intersecting at E   Given

2. AB ‖ CD   Opposite sides of a parallelogram are parallel

3. ∠BCD ≅ ∠ABC   If parallel lines are cut by a transversal, their alternate interior angles are congruent.

4. AB ≅ CD   Opposite sides of a parallelogram are congruent.

5. ∠BAD ≅ ∠ADC   If parallel lines are cut by a transversal, their alternate interior angles are congruent.

6. △ABE ≅ △CDE   ASA

7. AE ≅ DE   Corresponding parts of congruent triangles are congruent.

A quadrilateral is a parallelogram if any <u>one</u> of these are true:

1. One pair of opposite sides is both parallel and congruent.
2. Both pairs of opposite sides are congruent.
3. Both pairs of opposite angles are congruent.
4. The diagonals bisect each other.

<u>Example:</u>

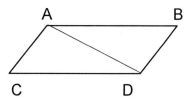

Given: Quadrilateral ABCD
AB $\cong$ CD
$\angle$BAC $\cong$ $\angle$ACD
Prove: ABCD is a parallelogram.

1. Quadrilateral ABCD
  AB $\cong$ CD
  $\angle$BAC $\cong$ $\angle$ACD
	Given

2. AD $\cong$ AD	Reflexive
3. $\triangle$ABD $\cong$ $\triangle$ACD	SAS
4. AC $\cong$ BD	Corresponding parts of congruent triangles are congruent.
5. ABCD is a parallelogram	If both pairs of opposite sides of a quadrilateral are congruent, the quadrilateral is a parallelogram.

A **rectangle** is a parallelogram with a right angle.

Rectangles have all the properties of parallelograms. In addition, all four angles are right angles and the diagonals are congruent.

A **rhombus** is a parallelogram with all sides equal length.

Rhombuses have all the properties of parallelograms. In addition, a rhombus is equilateral, the diagonals are perpendicular and the diagonals bisect the angles.

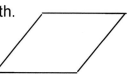

A **square** is a rectangle with all sides equal length.

Squares have all the properties of rectangles and all the properties of rhombuses.

A **trapezoid** is a quadrilateral with <u>exactly</u> one pair of opposite sides parallel. The parallel sides are called bases; the non-parallel sides are legs. In an **isosceles trapezoid,** the legs are congruent.

An **altitude** is a line segment drawn from a point on either base, perpendicular to the opposite base. The **median** is a line segment that joins the midpoints of each leg.

Theorems:

1. The median of a trapezoid is parallel to the bases and equal to one-half the sum of the lengths of the bases.

2. The base angles of an isosceles trapezoid are congruent.

3. The diagonals of an isosceles trapezoid are congruent.

Example:

In trapezoid ABCD, AB = 17 and CD = 21. Find the length of the median.

The median is one-half the sum of the bases.
$\frac{1}{2}(17 + 21) = 19$

Apply the distance formula.

The key to applying the distance formula is to understand the problem before beginning.

$$D = \sqrt{(x_2 - x_1)^2 + (y_2 - y_1)^2}$$

Sample Problem:

1. Find the perimeter of a figure with vertices at $(4,5)$, $(^-4,6)$ and $(^-5,^-8)$.

The figure being described is a triangle. Therefore, the distance for all three sides must be found. Carefully, identify all three sides before beginning.

Side $1 = (4,5)$ to $(^-4,6)$
Side $2 = (^-4,6)$ to $(^-5,^-8)$
Side $3 = (^-5,^-8)$ to $(4,5)$

$$D_1 = \sqrt{(^-4 - 4)^2 + (6 - 5)^2} = \sqrt{65}$$

$$D_2 = \sqrt{((^-5 - (^-4))^2 + (^-8 - 6)^2} = \sqrt{197}$$

$$D_3 = \sqrt{((4 - (^-5))^2 + (5 - (^-8))^2} = \sqrt{250} \text{ or } 5\sqrt{10}$$

$$\text{Perimeter} = \sqrt{65} + \sqrt{197} + 5\sqrt{10}$$

Apply the formula for midpoint.

Midpoint Definition:

If a line segment has endpoints of (x_1, y_1) and (x_2, y_2), then the midpoint can be found using:

$$\left(\frac{x_1 + x_2}{2}, \frac{y_1 + y_2}{2} \right)$$

Sample problems:

1. Find the center of a circle with a diameter whose endpoints are (3,7) and ($^-4$, $^-5$).

$$\text{Midpoint} = \left(\frac{3 + (^-4)}{2}, \frac{7 + (^-5)}{2} \right)$$

$$\text{Midpoint} = \left(\frac{^-1}{2}, 1 \right)$$

2. Find the midpoint given the two points $\left(5, 8\sqrt{6} \right)$ and $\left(9, ^-4\sqrt{6} \right)$.

$$\text{Midpoint} = \left(\frac{5 + 9}{2}, \frac{8\sqrt{6} + (^-4\sqrt{6})}{2} \right)$$

$$\text{Midpoint} = \left(7, 2\sqrt{6} \right)$$

Identify the coordinates of the vertices of a given polygon when it lies in the coordinate plane.

We can represent any two-dimensional geometric figure in the **Cartesian** or **rectangular coordinate system**. The Cartesian or rectangular coordinate system is formed by two perpendicular axes (coordinate axes): the X-axis and the Y-axis. If we know the dimensions of a two-dimensional, or planar, figure, we can use this coordinate system to visualize the shape of the figure.

Example: Represent an isosceles triangle with two sides of length 4.

Draw the two sides along the x- and y- axes and connect the points (vertices).

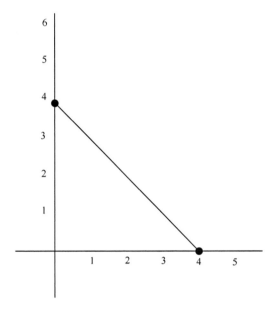

Identify the two-dimensional view of a three-dimensional object.

In order to represent three-dimensional figures, we need three coordinate axes (X, Y, and Z) which are all mutually perpendicular to each other. Since we cannot draw three mutually perpendicular axes on a two-dimensional surface, we use oblique representations.

Example: Represent a cube with sides of 2.

Once again, we draw three sides along the three axes to make things easier.

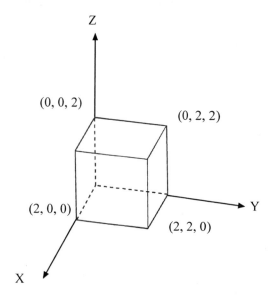

Each point has three coordinates (x, y, z).

SUBAREA V.   DATA ANALYSIS AND PROBABILITY

COMPETENCY 0012   UNDERSTAND METHODS OF COLLECTING, ORGANIZING, AND DESCRIBING DATA.

Identify appropriate graphical representations for a given data set.

BAR, LINE, PICTO-, AND CIRCLE GRAPHS

	Test 1	Test 2	Test 3	Test 4	Test 5
Evans, Tim	75	66	80	85	97
Miller, Julie	94	93	88	97	98
Thomas, Randy	81	86	88	87	90

Bar graphs are used to compare various quantities.

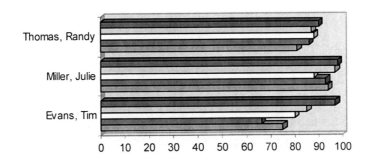

Line graphs are used to show trends, often over a period of time.

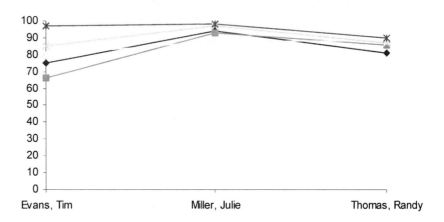

A **pictograph** shows comparison of quantities using symbols. Each symbol represents a number of items.

¶ ¶ ¶ ¶
¶ ¶ ¶ ¶ ¶ ¶ ¶
¶ ¶ ¶
¶ ¶
¶ ¶ ¶ ¶ ¶

Circle graphs show the relationship of various parts to each other and the whole. Percents are used to create circle graphs.

Julie spends 8 hours each day in school, 2 hours doing homework, 1 hour eating dinner, 2 hours watching television, 10 hours sleeping and the rest of the time doing other things.

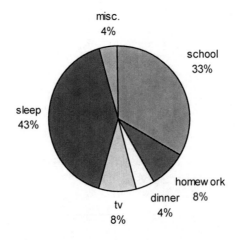

Compute the mean, median, mode, and range of a set of data.

Mean, median and mode are three measures of central tendency. The **mean** is the average of the data items. The **median** is found by putting the data items in order from smallest to largest and selecting the item in the middle (or the average of the two items in the middle). The **mode** is the most frequently occurring item.

Range is a measure of variability. It is found by subtracting the smallest value from the largest value.

Sample problem:

Find the mean, median, mode and range of the test score listed below:

85	77	65
92	90	54
88	85	70
75	80	69
85	88	60
72	74	95

Mean (X) = sum of all scores ÷ number of scores
       = 78

Median = put numbers in order from smallest to largest. Pick middle number.

54, 60, 65, 69, 70, 72, 74, 75, 77, 80, 85, 85, 85, 88, 88, 90, 92, 95
                              -- --
                         both in middle

Therefore, median is average of two numbers in the middle or 78.5

Mode = most frequent number
     = 85

Range = largest number minus the smallest number
      = 95 – 54
      = 41

Determine whether the mean, median, or mode is the most appropriate measure of central tendency in a given situation.

Different situations require different information. If we examine the circumstances under which an ice cream store owner may use statistics collected in the store, we find different uses for different information.

Over a 7-day period, the store owner collected data on the ice cream flavors sold. He found the mean number of scoops sold was 174 per day. The most frequently sold flavor was vanilla. This information was useful in determining how much ice cream to order in all and in what amounts for each flavor.

In the case of the ice cream store, the median and range had little business value for the owner.

Consider the set of test scores from a math class: 0, 16, 19, 65, 65, 65, 68, 69, 70, 72, 73, 73, 75, 78, 80, 85, 88, and 92. The mean is 64.06 and the median is 71. Since there are only three scores less than the mean out of the eighteen score, the median (71) would be a more descriptive score.

Retail store owners may be most concerned with the most common dress size so they may order more of that size than any other.

Probability measures the chances of an event occurring. The probability of an event that *must* occur, a certain event, is **one.** When no outcome is favorable, the probability of an impossible event is **zero**

$$P(event) = \frac{\text{number of favorable outcomes}}{\text{number of possible outcomes}}$$

Example: Given one die with faces numbered 1 - 6, the probability of tossing an even number on one throw of the die is 3/6 or ½ since there are 3 favorable outcomes (even faces) and a total of 6 possible outcomes (faces).

Example: If a fair die is rolled.

a)Find the probability of rolling an even number
b)Find the probability of rolling a number less than three.

a)The sample space is

S = {1, 2, 3, 4, 5, 6} and the event representing even numbers is

E = {2, 4, 6}

Hence, the probability of rolling an even number is

$$p(E) = \frac{n(E)}{n(S)} = \frac{3}{6} = \frac{1}{2} \text{ or } 0.5$$

b)The event of rolling a number less than three is represented by

A = {1, 2}

Hence, the probability of rolling a number less than three is

$$p(A) = \frac{n(A)}{n(S)} = \frac{2}{6} = \frac{1}{3} \text{ or } 0.33$$

Example: A class has thirty students. Out of the thirty students, twenty-four are males. Assuming all the students have the same chance of being selected, find the probability of selecting a female. (Only one person is selected.)

The number of females in the class is

30 - 24 = 6
Hence, the probability of selecting a female is

$$p(female) = \frac{6}{30} = \frac{1}{} \text{ or } 0.2$$

COMPETENCY 0013 UNDERSTAND THE THEORY AND APPLICATIONS OF PROBABILITY.

**Apply counting principles to solve real-world problems.
The Addition Principle of Counting** states:

If A and B are events, $n(A\,or\,B) = n(A) + n(B) - n(A \cap B)$.

Example:

In how many ways can you select a black card or a Jack from an ordinary deck of playing cards?

Let B denote the set of black cards and let J denote the set of Jacks. Then, $n(B) = 26, n(J) = 4, n(B \cap J) = 2$ and
$$n(B\,or\,J) = n(B) + n(J) - n(B \cap A)$$
$$= 26 + 4 - 2$$
$$= 28.$$

The Addition Principle of Counting for Mutually Exclusive Events states:

If A and B are mutually exclusive events, $n(A\,or\,B) = n(A) + n(B)$.

Example:

A travel agency offers 40 possible trips: 14 to Asia, 16 to Europe and 10 to South America. In how many ways can you select a trip to Asia or Europe through this agency?

Let A denote trips to Asia and let E denote trips to Europe. Then, $A \cap E = \varnothing$ and

$$n(A\,or\,E) = 14 + 16 = 30.$$

Therefore, the number of ways you can select a trip to Asia or Europe is 30.

The Multiplication Principle of Counting for Dependent Events states:

Let A be a set of outcomes of Stage 1 and B a set of outcomes of Stage 2. Then the number of ways $n(AandB)$, that A and B can occur in a two-stage experiment is given by:

$$n(AandB) = n(A)n(B|A),$$

where $n(B|A)$ denotes the number of ways B can occur given that A has already occurred.

Example:
How many ways from an ordinary deck of 52 cards can two Jacks be drawn in succession if the first card is drawn but not replaced in the deck and then the second card is drawn?

This is a two-stage experiment for which we wish to compute $n(AandB)$, where A is the set of outcomes for which a Jack is obtained on the first draw and B is the set of outcomes for which a Jack is obtained on the second draw.

If the first card drawn is a Jack, then there are only three remaining Jacks left to choose from on the second draw. Thus, drawing two cards without replacement means the events A and B are dependent.

$$n(AandB) = n(A)n(B|A) = 4 \cdot 3 = 12$$

The Multiplication Principle of Counting for Independent Events states:

Let A be a set of outcomes of Stage 1 and B a set of outcomes of Stage 2. If A and B are independent events then the number of ways $n(AandB)$, that A and B can occur in a two-stage experiment is given by:

$$n(AandB) = n(A)n(B).$$

Example:
How many six-letter code "words" can be formed if repetition of letters is not allowed?

Since these are code words, a word does not have to look like a word; for example, abcdef could be a code word. Since we must choose a first letter *and* a second letter *and* a third letter *and* a fourth letter *and* a fifth letter *and* a sixth letter, this experiment has six stages.

Since repetition is not allowed there are 26 choices for the first letter; 25 for the second; 24 for the third; 23 for the fourth; 22 for the fifth; and 21 for the sixth. Therefore, we have:

n(six-letter code words without repetition of letters)

$$= 26 \cdot 25 \cdot 24 \cdot 23 \cdot 22 \cdot 21$$

$$= 165,765,600$$

Determine probabilities of dependent or independent events.

Dependent events occur when the probability of the second event depends on the outcome of the first event. For example, consider the two events (A) it is sunny on Saturday and (B) you go to the beach. If you intend to go to the beach on Saturday, rain or shine, then A and B may be independent. If however, you plan to go to the beach only if it is sunny, then A and B may be dependent. In this situation, the probability of event B will change depending on the outcome of event A.

Suppose you have a pair of dice, one red and one green. If you roll a three on the red die and then roll a four on the green die, we can see that these events do not depend on the other. The total probability of the two independent events can be found by multiplying the separate probabilities.

$$P(A \text{ and } B) = P(A) \times P(B)$$
$$= 1/6 \times 1/6$$
$$= 1/36$$

Many times, however, events are not independent. Suppose a jar contains 12 red marbles and 8 blue marbles. If you randomly pick a red marble, replace it and then randomly pick again, the probability of picking a red marble the second time remains the same. However, if you pick a red marble, and then pick again without replacing the first red marble, the second pick becomes dependent upon the first pick.

$$P(\text{Red and Red}) \text{ with replacement} = P(\text{Red}) \times P(\text{Red})$$
$$= 12/20 \times 12/20$$
$$= 9/25$$
$$P(\text{Red and Red}) \text{ without replacement} = P(\text{Red}) \times P(\text{Red})$$
$$= 12/20 \times 11/19$$
$$= 33/95$$

Predict odds of a given outcome.

Odds are defined as the ratio of the number of favorable outcomes to the number of unfavorable outcomes. The sum of the favorable outcomes and the unfavorable outcomes should always equal the total possible outcomes.

For example, given a bag of 12 red and 7 green marbles compute the odds of randomly selecting a red marble.

$$\text{Odds of red} = \frac{12}{19}$$

$$\text{Odds of not getting red} = \frac{7}{19}$$

In the case of flipping a coin, it is equally likely that a head or a tail will be tossed. The odds of tossing a head are 1:1. This is called even odds.

Identify an appropriate sample space to determine the probability of a given event.

In probability, the **sample space** is a list of all possible outcomes of an experiment. For example, the sample space of tossing two coins is the set {HH, HT, TT, TH}, the sample space of rolling a six-sided die is the set {1, 2, 3, 4, 5, 6}, and the sample space of measuring the height of students in a class is the set of all real numbers {R}.

When conducting experiments with a large number of possible outcomes it is important to determine the size of the sample space. The size of the sample space can be determined by using the fundamental counting principle and the rules of combinations and permutations.

The **fundamental counting principle** states that if there are m possible outcomes for one task and n possible outcomes of another, there are
(m x n) possible outcomes of the two tasks together.

A **permutation** is the number of possible arrangements of items, without repetition, where order of selection is important.

A **combination** is the number of possible arrangements, without repetition, where order of selection is not important.

Examples:

1. Find the size of the sample space of rolling two six-sided die and flipping two coins.

   Solution:

   List the possible outcomes of each event:
   each dice: {1, 2, 3, 4, 5, 6}
   each coin: {Heads, Tails}

   Apply the fundamental counting principle:
   size of sample space = 6 x 6 x 2 x 2 = 144

Make predictions that are based on experimental or theoretical probabilities.

The absolute probability of some events cannot be determined. For instance, one cannot assume the probability of winning a tennis match is ½ because, in general, winning and losing are not equally likely. In such cases, past results of similar events can be used to help predict future outcomes. The **relative frequency** of an event is the number of times an event has occurred divided by the number of attempts.

$$\text{Relative frequency} = \frac{\text{number of successful trials}}{\text{total number of trials}}$$

For example, if a weighted coin flipped 50 times lands on heads 40 times and tails 10 times, the relative frequency of heads is 40/50 = 4/5. Thus, one can predict that if the coin is flipped 100 times, it will land on heads 80 times.

Example:

Two tennis players, John and David, have played each other 20 times.
John has won 15 of the previous matches and David has won 5.
(a) Estimate the probability that David will win the next match.
(b) Estimate the probability that John will win the next 3 matches.

Solution:

(a) David has won 5 out of 20 matches. Thus, the relative frequency of David winning is 5/20 or ¼. We can estimate that the probability of David winning the next match is ¼.

(b) John has won 15 out of 20 matches. The relative frequency of John winning is 15/20 or ¾. We can estimate that the probability of John winning a future match is ¾. Thus, the probability that John will win the next three matches is ¾ x ¾ x ¾ = 27/64.

COMPETENCY 0014   UNDERSTAND THE PROCESS OF ANALYZING AND INTERPRETING DATA TO MAKE STATISTICAL INFERENCES.

Random sampling is the process of studying an aspect of a population by selecting and gathering data from a segment of the population and making inferences and generalizations based on the results. Two main types of random sampling are simple and stratified. With simple random sampling, each member of the population has an equal chance of selection to the sample group. With stratified random sampling, each member of the population has a known but unequal chance of selection to the sample group, as the study selects a random sample from each population demographic. In general, stratified random sampling is more accurate because it provides a more representative sample group. Sample statistics are important generalizations about the entire sample such as mean, median, mode, range, and sampling error (standard deviation). Various factors affect the accuracy of sample statistics and the generalizations made from them about the larger population.

Sample size is one important factor in the accuracy and reliability of sample statistics. As sample size increases, sampling error (standard deviation) decreases. Sampling error is the main determinant of the size of the confidence interval. Confidence intervals decrease in size as sample size increases. A confidence interval gives an estimated range of values, which is likely to include a particular population parameter. The confidence level associated with a confidence interval is the probability that the interval contains the population parameter. For example, a poll reports 60% of a sample group prefers candidate A with a margin of error of $\pm$ 3% and a confidence level of 95%. In this poll, there is a 95% chance that the preference for candidate A in the whole population is between 57% and 63%.

The ultimate goal of sampling is to make generalizations about a population based on the characteristics of a random sample. Estimators are sample statistics used to make such generalizations. For example, the mean value of a sample is the estimator of the population mean. Unbiased estimators, on average, accurately predict the corresponding population characteristic. Biased estimators, on the other hand, do not exactly mirror the corresponding population characteristic. While most estimators contain some level of bias, limiting bias to achieve accurate projections is the goal of statisticians.

A **normal distribution** is the distribution associated with most sets of real-world data. It is frequently called a **bell curve**. A normal distribution has a **random variable** X with mean μ and variance σ^2.

Example:

Albert's Bagel Shop's morning customer load follows a normal distribution, with **mean** (average) 50 and **standard deviation** 10. The standard deviation is the measure of the variation in the distribution. Determine the probability that the number of customers tomorrow will be less than 42.

First convert the raw score to a **z-score**. A z-score is a measure of the distance in standard deviations of a sample from the mean.

The z-score = $\dfrac{X_i = \overline{X}}{s} = \dfrac{42-50}{10} = \dfrac{-8}{10} = -.8$

Next, use a table to find the probability corresponding to the z-score. The table gives us .2881. Since our raw score is negative, we subtract the table value from .5.

$$.5 - .2881 = .2119$$

We can conclude that $P(x < 42) = .2119$. This means that there is about a 21% chance that there will be fewer than 42 customers tomorrow morning.

Example:

The scores on Mr. Rogers' statistics exam follow a normal distribution with mean 85 and standard deviation 5. A student is wondering what the probability is that she will score between a 90 and a 95 on her exam.

We wish to compute $P(90 < x < 95)$.

Compute the z-scores for each raw score.

$$\frac{90-85}{5} = \frac{5}{5} = 1 \text{ and } \frac{95-85}{5} = \frac{10}{5} = 2.$$

Now we want $P(1 < z < 2)$.

Since we are looking for an occurrence between two values, we subtract:

$$P(1 < z < 2) = P(z < 2) - P(z < 1).$$

We use a table to get

$P(1 < z < 2) = .9772 - .8413 = .1359.$ (Remember that since the z-scores are positive, we add .5 to each probability.)

We can then conclude that there is a 13.6% chance that the student will score between a 90 and a 95 on her exam.

A **linear regression** equation is of the form: $Y = a + bX$.

Example:

A teacher wanted to determine how a practice test influenced a student's performance on the actual test. The practice test grade and the subsequent actual test grade for each student are given in the table below:

Practice Test (x)	Actual Test (y)
94	98
95	94
92	95
87	89
82	85
80	78
75	73
65	67
50	45
20	40

We determine the equation for the linear regression line to be $y = 14.650 + 0.834x$.

A new student comes into the class and scores 78 on the practice test. Based on the equation obtained above, what would the teacher predict this student would get on the actual test?

$$y = 14.650 + 0.834(78)$$
$$y = 14.650 + 65.052$$
$$y = 80$$

The law of large numbers and the central limit theorem are two fundamental concepts in statistics. The law of large numbers states that the larger the sample size, or the more times we measure a variable in a population, the closer the sample mean will be to the population mean. For example, the average weight of 40 apples out of a population of 100 will more closely approximate the population average weight than will a sample of 5 apples. The central limit theorem expands on the law of large numbers. The central limit theorem states that as the number of samples increases, the distribution of sample means (averages) approaches a normal distribution. This holds true regardless of the distribution of the population. Thus, as the number of samples taken increases the sample mean becomes closer to the population mean. This property of statistics allows us to analyze the properties of populations of unknown distribution.

In conclusion, the law of large numbers and central limit theorem show the importance of large sample size and large number of samples to the process of statistical inference. As sample size and the number of samples taken increase, the accuracy of conclusions about the population drawn from the sample data increases.

COMPETENCY 0015   UNDERSTAND HOW TO USE A VARIETY OF REPRESENTATIONS TO COMMUNICATE MATHEMATICAL IDEAS AND CONCEPTS AND CONNECTIONS BETWEEN THEM.

In order to understand mathematics and solve problems, one must know the definitions of basic mathematic terms and concepts. For a list of definitions and explanations of basic math terms, visit the following website: http://home.blarg.net/~math/deflist.html

Additionally, one must use the language of mathematics correctly and precisely to accurately communicate concepts and ideas.

For example, the statement "minus ten times minus five equals plus fifty" is incorrect because minus and plus are arithmetic operations not numerical modifiers. The statement should read "negative ten times negative five equals positive 50".

Symbolic representation is the basic language of mathematics. Converting data to symbols allows for easy manipulation and problem solving. Students should have the ability to recognize what the symbolic notation represents and convert information into symbolic form. For example, from the graph of a line, students should have the ability to determine the slope and intercepts and derive the line's equation from the observed data. Another possible application of symbolic representation is the formulation of algebraic expressions and relations from data presented in word problem form.

Examples, illustrations, and symbolic representations are useful tools in explaining and understanding mathematical concepts. The ability to create examples and alternative methods of expression allows students to solve real world problems and better communicate their thoughts.

Concrete examples are real world applications of mathematical concepts. For example, measuring the shadow produced by a tree or building is a real world application of trigonometric functions, acceleration or velocity of a car is an application of derivatives, and finding the volume or area of a swimming pool is a real world application of geometric principles.

Mathematical concepts and procedures can take many different forms. Students of mathematics must be able to recognize different forms of equivalent concepts.

For example, we can represent the slope of a line graphically, algebraically, verbally, and numerically. A line drawn on a coordinate plane will show the slope. In the equation of a line, $y = mx + b$, the term m represents the slope. We can define the slope of a line several different ways. The slope of a line is the change in the value of the y divided by the change in the value of x over a given interval. Alternatively, the slope of a line is the ratio of "rise" to "run" between two points. Finally, we can calculate the numeric value of the slope by using the verbal definitions and the algebraic representation of the line.

Identify appropriate representations or models for mathematics operations or situations using written, concrete, pictorial, graphical, or algebraic methods.

Mathematical operations include addition, subtraction, multiplication and division.

Addition can be indicated by the expressions: sum, greater than, and, more than, increased by, added to.

Subtraction can be expressed by: difference, fewer than, minus, less than, decreased by.

Multiplication is shown by: product, times, multiplied by, twice.

Division is used for: quotient, divided by, ratio.

Examples:
7 added to a number	$n + 7$
a number decreased by 8	$n - 8$
12 times a number divided by 7	$12n \div 7$
28 less than a number	$n - 28$
the ratio of a number to 55	$\dfrac{n}{55}$
4 times the sum of a number and 21	$4(n + 21)$

Mathematical operations can be shown using manipulatives or drawings.

Multiplication can be shown using arrays.

3×4

Addition and subtractions can be demonstrated with symbols.

ψ ψ ψ ξ ξ ξ
$3 + 4 = 7$
$7 - 3 = 4$

Fractions can be clarified using pattern blocks, fraction bars, or paper folding.

MANIPULATIVES

Example:
Using tiles to demonstrate both geometric ideas and number theory.

Give each group of students 12 tiles and instruct them to build rectangles. Students draw their rectangles on paper.

12×1

1×12

3×4

Students of mathematics must be able to recognize and interpret the different representations of arithmetic operations.

First, there are many different verbal descriptions for the operations of addition, subtraction, multiplication, and division. The table below identifies several words and/or phrases that are often used to denote the different arithmetic operations.

Operation	Descriptive Words
Addition	"plus", "combine", "sum", "total", "put together"
Subtraction	"minus", "less", "take away", "difference"
Multiplication	"product", "times", "groups of"
Division	"quotient", "into", "split into equal groups",

Second, diagrams of arithmetic operations can present mathematical data in visual form. For example, we can use the number line to add and subtract.

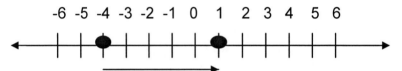

The addition of 5 to -4 on the number line; $-4 + 5 = 1$.

Finally, as shown in the examples below, we can use pictorial representations to explain all of the arithmetic processes.

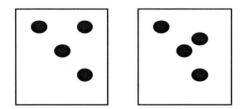

Two groups of four equals eight or $2 \times 4 = 8$ shown in picture form.

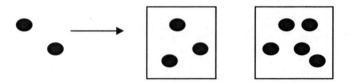

Adding three objects to two or $3 + 2 = 5$ shown in picture form.

Identify appropriate mathematical problems from real-world situations.

The unit rate for purchasing an item is its price divided by the number of pounds/ ounces, etc. in the item. The item with the lower unit rate is the lower price.

Example: Find the item with the best unit price:

$1.79 for 10 ounces
$1.89 for 12 ounces
$5.49 for 32 ounces

$$\frac{1.79}{10} = .179 \text{ per ounce} \qquad \frac{1.89}{12} = .1575 \text{ per ounce} \qquad \frac{5.49}{32} = .172 \text{ per ounce}$$

$1.89 for 12 ounces is the best price.

A second way to find the better buy is to make a proportion with the price over the number of ounces, etc. Cross multiply the proportion, writing the products above the numerator that is used. The better price will have the smaller product.

Example: Find the better buy:

$8.19 for 40 pounds or $4.89 for 22 pounds

Find the unit price.

$$\frac{40}{8.19} = \frac{1}{x} \qquad\qquad \frac{22}{4.89} = \frac{1}{x}$$

$$40x = 8.19 \qquad\qquad 22x = 4.89$$

$$x = .20475 \qquad\qquad x = .22\overline{227}$$

Since $.20475 < .22\overline{227}$, $8.19 is less and is a better buy.

Use mathematics to solve problems in other contexts.

Elapsed time problems are usually one of two types. One type of problem is the elapsed time between 2 times given in hours, minutes, and seconds. The other common type of problem is between 2 times given in months and years.

For any time of day past noon, change it into military time by adding 12 hours. For instance, 1:15 p.m. would be 13:15. Remember when you borrow a minute or an hour in a subtraction problem that you have borrowed 60 more seconds or minutes.

Example: Find the time from 11:34:22 a.m. until 3:28:40 p.m.

First change 3:28:40 p.m. to   15:28:40 p.m.
Now subtract   - 11:34:22 a.m.
                                                       :18

Borrow an hour and add 60 more minutes. Subtract
14:88:40 p.m.
- 11:34:22 a.m.
   3:54:18 ↔ 3 hours, 54 minutes, 18 seconds

Example: John lived in Arizona from September 91 until March 95. How long is that?

		year	month
March 95	=	95	03
September 91	= -	91	09

Borrow a year, change it into 12 more months, and **subtract.**

		year	month
March 95	=	94	15
September 91	= -	91	09
		3 yr	6 months

Example: A race took the winner 1 hr. 58 min. 12 sec. on the first half of the race and 2 hr. 9 min. 57 sec. on the second half of the race. How much time did the entire race take?

```
  1 hr. 58 min. 12 sec.
+ 2 hr.  9 min. 57 sec.     Add these numbers
  3 hr. 67 min. 69 sec.
+ 1 min -60 sec.           Change 60 seconds to 1 min.
  3 hr. 68 min.  9 sec.
+ 1 hr.-60 min.            Change 60 minutes to 1 hr.
  4 hr.  8 min.  9 sec.  ← final answer
```

COMPETENCY 0016   UNDERSTAND MATHEMATICAL REASONING, THE CONSTRUCTION OF MATHEMATICAL ARGUMENTS, AND PROBLEM-SOLVING STRATEGIES IN MATHEMATICS AND OTHER CONTEXTS.

Inductive thinking is the process of finding a pattern from a group of examples. That pattern is the conclusion that this set of examples seemed to indicate. It may be a correct conclusion or it may be an incorrect conclusion because other examples may not follow the predicted pattern.

Deductive reasoning is the process of arriving at a conclusion based on other statements that are all known to be true.

A symbolic argument consists of a set of premises and a conclusion in the format of if [Premise 1 and premise 2] then [conclusion].

An argument is **valid** when the conclusion follows necessarily from the premises.   An argument is **invalid** or a fallacy when the conclusion does not follow from the premises.

There are 4 standard forms of valid arguments which must be remembered.

1. Law of Detachment	If p, then q	(premise 1)
	p,	(premise 2)
	Therefore, q	
2. Law of Contraposition	If p, then q	
	not q,	
	Therefore not p	
3. Law of Syllogism	If p, then q	
	If q, then r	
	Therefore if p, then r	
4. Disjunctive Syllogism	p or q	
	not p	
	Therefore, q	

A counterexample is an exception to a proposed rule or conjecture that disproves the conjecture. For example, the existence of a single non-brown dog disproves the conjecture "all dogs are brown". Thus, any non-brown dog is a counterexample.

In searching for mathematic counterexamples, one should consider extreme cases near the ends of the domain of an experiment and special cases where an additional property is introduced. Examples of extreme cases are numbers near zero and obtuse triangles that are nearly flat. An example of a special case for a problem involving rectangles is a square because a square is a rectangle with the additional property of symmetry.

Example:

Identify a counterexample for the following conjectures.

1. If n is an even number, then $n + 1$ is divisible by 3.

> $n = 4$
> $n + 1 = 4 + 1 = 5$
> 5 is not divisible by 3.

2. If n is divisible by 3, then $n^2 - 1$ is divisible by 4.

> $n = 6$
> $n^2 - 1 = 6^2 - 1 = 35$
> 35 is not divisible by 4.

Identify valid mathematical arguments (e.g., an explanation that the sum of two odd numbers is always even).

A valid argument is a statement made about a pattern or relationship between elements, thought to be true, which is subsequently justified through repeated examples and logical reasoning. Another term for a valid argument is a proof.

For example, the statement that the sum of two odd numbers is always even could be tested through actual examples.

Two Odd Numbers	Sum	Validity of Statement
1+1	2 (even)	Valid
1+3	4 (even)	Valid
61+29	90 (even)	Valid
135+47	182 (even)	Valid
253+17	270 (even)	Valid
1,945+2,007	3,952 (even)	Valid
6,321+7,851	14,172 (even)	Valid

Adding two odd numbers always results in a sum that is even. It is a valid argument based on the justifications in the table above.

Here is another example. The statement that a fraction of a fraction can be determined by multiplying the numerator by the numerator and the denominator by the denominator can be proven through logical reasoning. For example, one-half of one-quarter of a candy bar can be found by multiplying ½ * ¼. The answer would be one-eighth. The validity of this argument can be demonstrated as valid with a model.

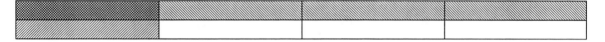

The entire rectangle represents one whole candy bar. The top half section of the model is shaded in one direction to demonstrate how much of the candy bar remains from the whole candy bar. The left quarter, shaded in a different direction, demonstrates that one-quarter of the candy bar has been given to a friend. Since the whole candy bar is not available to give out, the area that is double-shaded is the fractional part of the ½ candy bar that has been actually given away. That fractional part is one-eighth of the whole candy bar as shown in both the sketch and the algorithm.

Successful math teachers introduce their students to multiple problem solving strategies and create a classroom environment where free thought and experimentation are encouraged. Teachers can promote problem solving by allowing multiple attempts at problems, giving credit for reworking test or homework problems, and encouraging the sharing of ideas through class discussion. There are several specific problem solving skills with which teachers should be familiar.

The **guess-and-check** strategy calls for students to make an initial guess at the solution, check the answer, and use the outcome of to guide the next guess. With each successive guess, the student should get closer to the correct answer. Constructing a table from the guesses can help organize the data.

Example:

There are 100 coins in a jar. 10 are dimes. The rest are pennies and nickels. There are twice as many pennies as nickels. How many pennies and nickels are in the jar?

There are 90 total nickels and pennies in the jar (100 coins – 10 dimes).

There are twice as many pennies as nickels. Make guesses that fulfill the criteria and adjust based on the answer found. Continue until we find the correct answer, 60 pennies and 30 nickels.

Number of Pennies	Number of Nickels	Total Number of Pennies and Nickels
40	20	60
80	40	120
70	35	105
60	30	90

When solving a problem where the final result and the steps to reach the result are given, students must **work backwards** to determine what the starting point must have been.

Example:

John subtracted seven from his age, and divided the result by 3. The final result was 4. What is John's age?

> Work backward by reversing the operations.
> 4 x 3 = 12;
> 12 + 7 = 19
> John is 19 years old.

Estimation and testing for **reasonableness** are related skills students should employ prior to and after solving a problem. These skills are particularly important when students use calculators to find answers.

Example:

Find the sum of 4387 + 7226 + 5893.

4300 + 7200 + 5800 = 17300	Estimation.
4387 + 7226 + 5893 = 17506	Actual sum.

By comparing the estimate to the actual sum, students can determine that their answer is reasonable.

Artists, musicians, scientists, social scientists, and business people use mathematical modeling to solve problems in their disciplines. These disciplines rely on the tools and symbology of mathematics to model natural events and manipulate data. Mathematics is a key aspect of visual art.

Artists use the geometric properties of shapes, ratios, and proportions in creating paintings and sculptures. For example, mathematics is essential to the concept of perspective. Artists must determine the appropriate lengths and heights of objects to portray three-dimensional distance in two dimensions.

Mathematics is also an important part of music. Many musical terms have mathematical connections. For example, the musical octave contains twelve notes and spans a factor of two in frequency. In other words, the frequency, the speed of vibration that determines tone and sound quality, doubles from the first note in an octave to the last. Thus, starting from any note we can determine the frequency of any other note with the following formula.

$$\text{Freq} = \text{note} \times 2^{N/12}$$

Where N is the number of notes from the starting point and note is the frequency of the starting note. Mathematical understanding of frequency plays an important role in the tuning of musical instruments.

In addition to the visual and auditory arts, mathematics is an integral part of most scientific disciplines. The uses of mathematics in science are almost endless. The following are but a few examples of how scientists use mathematics. Physical scientists use vectors, functions, derivatives, and integrals to describe and model the movement of objects. Biologists and ecologists use mathematics to model ecosystems and study DNA. Finally, chemists use mathematics to study the interaction of molecules and to determine proper amounts and proportions of reactants.

Many social science disciplines use mathematics to model and solve problems. Economists, for example, use functions, graphs, and matrices to model the activities of producers, consumers, and firms. Political scientists use mathematics to model the behavior and opinions of the electorate. Finally, sociologists use mathematical functions to model the behavior of humans and human populations.

Finally, mathematical problem solving and modeling is essential to business planning and execution. For example, businesses rely on mathematical projections to plan business strategy. Additionally, stock market analysis and accounting rely on mathematical concepts.

Mathematical concepts and procedures can take many different forms. Students of mathematics must be able to recognize different forms of equivalent concepts.

For example, we can represent the slope of a line graphically, algebraically, verbally, and numerically. A line drawn on a coordinate plane will show the slope. In the equation of a line, $y = mx + b$, the term m represents the slope. We can define the slope of a line several different ways. The slope of a line is the change in the value of the y divided by the change in the value of x over a given interval. Alternatively, the slope of a line is the ratio of "rise" to "run" between two points. Finally, we can calculate the numeric value of the slope by using the verbal definitions and the algebraic representation of the line.

There are many different ways to solve mathematical problems other than manual calculation. Although we are all familiar with the use of calculators to solve simple numerical problems, innovations in computer software have permitted the development of programs that allow computers to assist in the exploration of ever more complicated concepts.

One of the most powerful uses of computers to aid in the comprehension of mathematics is the use of programs to visualize mathematical concepts and functions. Function graphs and dynamic geometry software can be used to visualize and manipulate equations in two and three dimensions. This can make it easier to understand concepts such as adding two functions together, algebraic transformations, and vector and matrix calculations. For many students, a graphical representation of mathematic problems makes them easier to comprehend. These types of software allow students to create and manipulate math in a concrete visual form that may be easier to graph than purely symbolic numerical calculations.

Another use for computing tools is the verification of solutions that were arrived at using other methods. Equations solved algebraically can frequently be verified graphically. Various types of programs can also be used to confirm the results of almost every type of basic calculation. When used as an adjunct to manual calculation techniques, this use of computing tools can help build students' confidence in their mathematical abilities.

Computers are also extremely useful in situations that involve multiple repeated calculations. For example, curve fitting software uses pre-programmed algorithms to find the best-fit curve for a given set of data, a process that is extremely cumbersome if done by hand. Such tools still require the user to determine what type of curve is most appropriate to use as a model for any given data set, but once the shape of that curve has been identified, it is much easier for the computer to do the grunt work.

Finally, symbolic manipulators, are extremely useful when solving and checking the results of complex algebraic problems whose solutions might otherwise be difficult to verify or understand. These programs can be used for such diverse functions as solving complex systems of equations, doing matrix and vector calculations, differentiating or integrating equations, and graphing complicated functions. In general, symbolic manipulators use their own programming languages in combination with preprogrammed functions to simplify the process of solving advanced algebraic problems. Their output, particularly the graphical representation of problems, can also make it easier for both students and teachers to communicate mathematical information.

CURRICULUM AND INSTRUCTION

TEACHING METHODS - The art and science specific for high school mathematics

Some commonly used teaching techniques and tools are described below along with links to further information. The links provided provide a wealth of instructional ideas and materials. You should consider joining <u>The National Council of Teachers of Mathematics</u> as they have many ideas in their journals about pedagogy and curriculum standards and publish professional books that are useful. You can write to them at 1906 Association Drive, Reston, VA 20191-1593. You can also order a starter kit from them for $9 that includes 3 recent journals by calling 800-235-7566 or writing <u>e-mailorders@nctm.org</u>.

> A couple of resources for students to use at home:
> <u>http://www.algebra.com/</u>, <u>http://www.mathsisfun.com/algebra/index.html</u> and <u>http://www.purplemath.com/</u>. A helpful .pdf guide for parents is available at:
> <u>http://my.nctm.org/ebusiness/ProductCatalog/product.aspx?ID=12931</u>
> A good website for understanding the causes of and how to prevent "math anxiety:"
> <u>http://www.mathgoodies.com/articles/math_anxiety.html</u>

1. Classroom warm-up: Engage your students as soon as they walk in the door: provide an interesting short activity each day. You can make use of thought-provoking questions and puzzles. Also use relevant puzzles specific to topics you may be covering in your class. The following websites provide some ideas:
   > <u>http://www.math-drills.com/?gclid=CP-P0dzenJICFRSTGgodNjG0Zw</u>
   > <u>http://www.mathgoodies.com/games/</u>
   > <u>http://mathforum.org/k12/k12puzzles/</u>
   > <u>http://mathforum.org/pow/other.html</u>

2. Real life examples: Connect math to other aspects of your students' lives by using examples and data from the real world whenever possible. It will not only keep them engaged, it will also help answer the perennial question "Why do we have to learn math?" Online resources to get you started:

http://chance.dartmouth.edu/chancewiki/index.php/Main_Page has some interesting real-world probability problems (such as, "Can statistics determine if Robert Clemens used steroids?")

http://www.mathnotes.com/nos_index.html has all kinds of links between math and the real world suitable for high school students

http://www.nssl.noaa.gov/edu/ideas/ uses weather to teach math

http://standards.nctm.org/document/eexamples/index.htm#9-12 Using Graphs, Equations, and Tables to Investigate the Elimination of Medicine from the Body: Modeling the Situation

http://mathforum.org/t2t/faq/election.html Election math in the classroom

http://www.education-world.com/a_curr/curr148.shtml offers examples of real-life problems such as calculating car payments, saving and investing, the world of credit cards, and other finance problems.

http://score.kings.k12.ca.us/real.world.html is a website connecting math to real jobs, elections, NASA projects, etc.

3. Graphing and spreadsheets for enhancing math learning:

http://www.cvgs.k12.va.us/digstats/

http://score.kings.k12.ca.us/standards/probability.html for graphing and statistics.

http://www.microsoft.com/education/solving.mspx for using spreadsheets and to solve polynomial problems.

4. Use technology - manipulatives, software and interactive online activities that can help all students learn, particularly those oriented more towards visual and kinesthetic learning. Here are some websites:

http://illuminations.nctm.org/ActivitySearch.aspx has games for grades 9-12 that can be played against the computer or another student.

http://nlvm.usu.edu/ The National Library of Virtual Manipulatives has resources for all grades on numbers and operations, algebra, geometry, probability and measurement.

http://mathforum.org/pow/other.html has links to various math challenges, manipulatives and puzzles.

http://www.etacuisenaire.com/algeblocks/algeblocks.jsp Algeblocks are blocks that utilize the relationship between algebra and geometry.

5. Word problem strategies- the hardest thing to do is take the English and turn it into math but there are 6 key steps to teach students how to solve word problems:

   a. The problem will have **key words** to suggest the type of operation or operations to be performed to solve the problem. For example, words such as "altogether" or "total" imply addition while words such as "difference" or "How many more?" imply subtraction.

   b. *Pictures or Concrete Materials*: Math is very abstract; it is easier to solve a problem using pictures or concrete materials to illustrate the problem. Pictures and concrete materials allow the students to manipulate the material to solve the problem with trial and error. Model drawing pictures and using concrete materials to solve word problems.

   c. *Use Logic*: Ask your students if their answers make sense. Get them used to using the process of deduction. Model the deduction process for them to decide on the answer to a word problem. For example, in solving a problem such as: Two consecutive numbers have a sum of 91. What are the numbers? If the student arrives at an answer of 44 and 45 it is obvious that there was an error in the equation used or calculation since 44 and 45 are consecutive but don't add up to 91. Let x = the 1st number and $(x+1)$ = the 2nd number, so that $x + (x+1)$ $=91$ and $2x +1 =91$, then $2x=90$, $x=45$ and $x+1=46$. The answer is 45 and 46.

   d. *Eliminate the possibilities and look for patterns or work the problem backwards*

   e. *Guess the Answer*: Students should guess an approximate answer that makes sense based on the problem. For example, if the student knows the word problem implies addition, they should recognize that the answer must be greater than the numbers in the problem. Often students are afraid of guessing because they don't want to get the wrong answer but encourage your students to guess and then double check the answer to see if it works. If the answer is incorrect, the student can try another strategy for finding the answer.

   f. *Make a Table*: Selecting relevant information from a word problem and organizing the data is very helpful in solving word problems. Often students become confused because there are too many numbers and/or variables.

http://www.purplemath.com/modules/translat.htm, http://math.about.com/library/weekly/aa071002a.htm and http://www.onlinemathlearning.com/algebra-word-problems.html are great resources for students to solve word problems.

6. Mental math practice

Give students regular practice in doing mental math. The following websites offer many mental calculation tips and strategies:
http://www.cramweb.com/math/index.htm
http://mathforum.org/k12/mathtips/mathtips.html

Because frequent calculator use tends to deprive students of a sense of numbers and an ability to calculate on their own, they will often approach a sequence of multiplications and divisions the hard way. For instance, asked to calculate 770 x 36/ 55, they will first multiply 770 and 36 and then do a long division with the 55. They fail to recognize that both 770 and 55 can be divided by 11 and then by 5 to considerably simplify the problem. Give students plenty of practice in multiplying and dividing a sequence of integers and fractions so they are comfortable with canceling top and bottom terms.

7. Math language

Math vocabulary help is available for high school students on the web:
http://www.amathsdictionaryforkids.com/ is a colorful website math dictionary
http://www.math.com/tables/index.html is a math dictionary in English and Spanish

ERROR ANALYSIS

A simple method for analyzing student errors is to ask how the answer was obtained. The teacher can then determine if a common error pattern has resulted in the wrong answer. There is a value to having the students explain how they arrived at the correct as well as the incorrect answers.

Many errors are due to simple **carelessness**. Students need to be encouraged to work slowly and carefully. They should check their calculations by redoing the problem on another paper, not merely looking at the work. Addition and subtraction problems need to be written neatly so the numbers line up. Students need to be careful regrouping in subtraction. Students must write clearly and legibly, including erasing fully. Use estimation to ensure that answers make sense.

Many students' computational skills exceed their **reading** level. Although they can understand basic operations, they fail to grasp the concept or completely understand the question. Students must read directions slowly.
Fractions are often a source of many errors. Students need to be reminded to use common denominators when adding and subtracting and to always express answers in simplest terms. Again, it is helpful to check by estimating.

The most common error that is made when working with **decimals** is failure to line up the decimal points when adding or subtracting or not moving the decimal point when multiplying or dividing. Students also need to be reminded to add zeroes when necessary. Reading aloud may also be beneficial. Estimation, as always, is especially important.

Students need to know that it is okay to make mistakes. The teacher must keep a positive attitude, so they do not feel defeated or frustrated.

THE ART OF TEACHING - PEDAGOGICAL PRINCIPLES
Maintain a supportive, non-threatening environment

The key to success in teaching goes beyond your mathematical knowledge and the desire to teach. Though important, knowledge and desire alone do not make you a good teacher. Being able to connect with your students is vital: learn their names immediately, have a seating chart the first day (even if you intend to change it) and learn about your students -their hobbies, phone number, parent's names, what they like and dislike about school and learning and math. Keep this information on each student _and learn it_: adapt your lessons, how challenging they are and what other resources you may need to accommodate your students' individual strengths and weaknesses.

Learn to see math as your students see it. If you aren't able to connect with your students, no matter how well your lessons are and how well you know the material, you won't inspire them to learn math from you. As you expect respect, you must give respect and as you expect their attention, they also need your attention and understanding. Talk to them with the same tone of voice as you would an adult, not in a tone that makes them feel like children. Look your students in the eye when you talk to them and encourage questions and comments. Take advantage of teachable moments and explain the rationale behind math rules.

Demonstrate respect, care and trust toward every student; assume the best. This does not mean becoming "friends" with your students or you will have problems with discipline. You can be kind and firm at the same time. Have a fair and clear grading and discipline system that is posted, reviewed and made clear to your students. Consistency, structure and fairness are essential to earning their trust in you as a teacher. Always admit your mistakes and be available to your students certain days after school. Finally, demonstrate your care for them and your love of math and you will be a positive influence on their learning.

Below are websites to help make your teaching more effective and fun:

1. **Teachers Helping Teachers** has several resources for high school mathematics.

2. **Math Resources for Teachers** – resources for grades 7 - 10
3. **Math is Marvelous Web Site** - is a fascinating website on the history of geometry
4. **Math Archives K-12** resources for lesson plans and software
5. **http://www.edhelper.com/algebra.htm** covers Algebra I & II
6. **Math Goodies** interactive lessons, worksheets and homework help
7. **Multicultural Lessons** an interesting site with lessons on Babylonian Square Roots, Chinese Fraction Reducing, Egyptian multiplication, etc.
8. **http://www.goenc.com/** resources and professional development for teachers
9. **Math and Reading Help** a guide to math, reading, homework help, tutoring and earning a high school diploma
10. **Purple Math.com** all about Algebra, lessons, help for students and lots of other resources
11. **http://library.thinkquest.org/20991/home.html** this site has Algebra, Geometry and Pre-calc/Calculus
12. **http://www.math.com/** this site has Algebra, Geometry, Trigonometry, and Calculus, plus homework help
13. **http://www.math.armstrong.edu/MathTutorial/index.html** a tutorial in algebra
14. **http://www.wtamu.edu/academic/anns/mps/math/mathlab/beg_algebra/index.htm** this site is helpful for those beginning Algebra or for a refresher
15. **Math Complete** this radicals, quadratics, linear equation solvers
16. **Math Tutor - PEMDAS & Integers** fractions, integers, information for parents and teachers
17. **Matrix Algebra** all about matrix operations and applications
18. **Mr. Stroh's Algebra Site** help for Algebra I & II
19. **Polynomials and Polynomial Functions** everything from factoring, to graphing, finding rational zeros and multiplying, adding and subtracting polynomials
20. **Quadratic Formula** all about quadratics
21. **Animated Pythagorean Theorem** a fun an animated proof of the Pythagorean Theorem
22. **Brunnermath - Interactive Activities** general math, Algebra, Geometry, Trigonometry, Statistics, Calculus, using Calculators
23. **CoolMath4Kids Geometry** creating art with math and geometry lessons
24. **The Curlicue Fractal** The curlicue fractal is an exceedingly easy-to-make but richly complex pattern using trigonometry and calculus to create fascinating shapes
25. **Euclid's Elements Interactive** Euclid's *Elements* form one of the most beautiful and influential works of science in the history of humankind.
26. **Howe-Two Free Software** software solutions for mathematics instruction
27. **http://regentsprep.org/regents/math/math-topic.cfm?TopicCode=syslin** Systems of equations lessons and practice

28. **http://www.sparknotes.com/math/algebra1/systemsofequations/problems3.rhtml** Word problems system of equations
29. **http://math.about.com/od/complexnumbers/Complex_Numbers.htm** Several complex number exercise pages
30. **http://regentsprep.org/Regents/math/ALGEBRA/AE3/PracWord.htm** practice with Systems of inequalities word problems
31. **http://regentsprep.org/regents/Math/solvin/PSolvIn.htm** solving inequalities
32. **http://www.wtamu.edu/academic/anns/mps/math/mathlab/beg_algebra/beg_alg_tut18_ineq.htm** Inequality tutorial, examples, problems
33. **http://www.wtamu.edu/academic/anns/mps/math/mathlab/beg_algebra/beg_alg_tut24_ineq.htm** Graphing linear inequalities tutorial
34. **http://www.wtamu.edu/academic/anns/mps/math/mathlab/col_algebra/col_alg_tut17_quad.htm** Quadratic equations tutorial, examples, problems
35. **http://regentsprep.org/Regents/math/math-topic.cfm?TopicCode=factor** Practice factoring
36. **http://www.wtamu.edu/academic/anns/mps/math/mathlab/col_algebra/col_alg_tut37_syndiv.htm** Synthetic division tutorial
37. **http://www.tpub.com/math1/10h.htm** Synthetic division Examples and problems

DEVELOPMENTAL PSYCHOLOGY AND TEACHING MATHEMATICS- things you may not know about your students:

Studies show that health matters more than gender or social status when it comes to learning. Healthy girls and boys do equally well on most cognitive tasks. Boys perform better at manipulating shapes and analyzing and girls perform better on processing speed and motor dexterity. No differences have been measured in calculation ability, meaning girls and boys have an equal aptitude for mathematics.

The following was written by Jay Giedd, M.D. is a practicing Child and Adolescent Psychiatrist and Chief of Brain Imaging at the Child Psychiatry Branch of the National Institute of Mental Health:

http://nihrecord.od.nih.gov/newsletters/2005/08_12_2005/story04.htm

"The most surprising thing has been how much the teen brain is changing. By age six, the brain is already 95 percent of its adult size. But the gray matter, or thinking part of the brain, continues to thicken throughout childhood as the brain cells get extra connections, much like a tree growing extra branches, twigs and roots...In the frontal part of the brain, the part of the brain involved in judgment, organization, planning, strategizing -- those very skills that teens get better and better at -- this process of thickening of the gray matter peaks at about age 11 in girls and age 12 in boys, roughly about the same time as puberty. After that peak, the gray matter thins as the excess connections are eliminated or pruned...

Contrary to what most parents have thought at least once, "teens really do have brains," quipped Dr. Jay Giedd, NIMH intramural scientist, in a lecture on the "Teen Brain Under Construction." His talk was the kick-off event for the recent NIH Parenting Festival. Giedd said scientists have only recently learned more about the trajectories of brain growth. One of the findings he discussed showed the frontal cortex area — which governs judgment, decision-making and impulse control — doesn't fully mature until around age 25. "That really threw us," he said. "We used to joke about having to be 25 to rent a car, but there's tons of data from insurance reports [showing] that 24-year-olds are costing them more than 44-year-olds."

So why is that? "It must be behavior and impulse control," he said. "Whatever these changes are, the top 10 bad things that happen to teens involve emotion and behavior." Physically, Giedd said, the teen years and early 20s represent an incredibly healthy time of life, in terms of cancer, heart disease and other serious illnesses. But with accidents as the leading cause of death in adolescents, and suicide following close behind, "this isn't a great time emotionally and psychologically. This is the great paradox of adolescence: right at the time you should be on the top of your game, you're not."

The next step in Giedd's research, he said, is to learn more about what influences brain growth, for good or bad. "Ultimately, we want to use these findings to treat illness and enhance development."

One of the things scientists have come to understand, though, is that parents do have something to do with their children's brain development.

"From imaging studies, one of the things that seems intriguing is this notion of modeling...that the brain is pretty adept at learning by example," he said. "As parents, we teach a lot when we don't even know we're teaching, just by showing how we treat our spouses, how we treat other people, what we talk about in the car on the way home...things that a parent says in the car can stick with them for years. They're listening even though it may appear they're not."

What can we do to change our kids? "Well, start with yourself in terms of what you show by example," Giedd concluded.

Maybe the parts of the brain performing geometry are different from the parts doing algebra. There is no definitive research to answer that question yet, but it is obviously what researchers are looking for.

Time-Lapse Imaging Tracks Brain Maturation from ages 5 to 20
Constructed from MRI scans of healthy children and teens, the time-lapse "movie", from which the above images were extracted, compresses 15 years of brain development (ages 5–20) into just a few seconds.

To view in color, go to the website below: Red indicates **more** gray matter, **blue less** gray matter. Gray matter wanes in a back-to-front wave as the brain matures and neural connections are pruned.

Source: Paul Thompson, Ph.D. UCLA Laboratory of Neuroimaging
http://www.loni.ucla.edu/%7Ethompson/DEVEL/PR.html

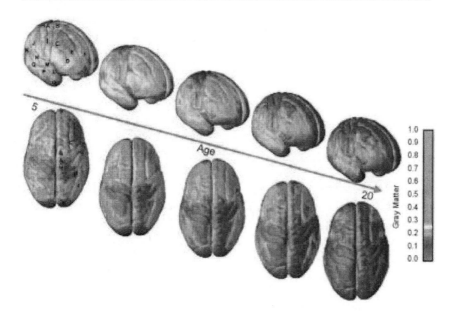

What are the implications of this fascinating study for teachers? It's unreasonable to expect teens to have adult levels of organizational skills or decision-making before their brains have completely developed. In teens, the frontal lobe, or the executive of the brain is what handles organizing, decision making, emotions, attending, shifting attention, planning and making strategies and it is not fully developed until the early to mid-twenties.

*Perhaps since certain parts of the brain develop sooner than others, subjects should be taught in a different order. Until we know more, just understanding that the parts of teen brains related to decision making and emotions are still developing through the early 20's is important, and **that stressful situations lead to diminished ability to made good judgments.*** For some children, just being called on in class is stressful. *At this age, social relationships become very important and **teachers need to be sensitive to this aspect of teen development as it relates to stress and decision-making.*** *The immaturity of this part of the teen brain might explain why the teen crash rate is **4** times that of adults...*

Answer Key to Practice Problems

Competency 0003, *page 13*

Question #1   x-intercept = -14   y-intercept = -10   slope = $-\dfrac{5}{7}$

Question #2   x-intercept = 14   y-intercept = -7   slope = $\dfrac{1}{2}$

Question #3   x-intercept = 3   y-intercept = none

Question #4   x-intercept = $\dfrac{15}{2}$ y-intercept = 3   slope = $-\dfrac{2}{5}$

page 18

Question #1   Domain = $^-\infty,\infty$ Range = $^-5,\infty$

Question #2   Domain = $^-\infty,\infty$ Range = $^-6,\infty$

Question #3   Domain = 1,4,7,6   Range = -2

Question #4   Domain = $x \neq 2, ^-2$

Question #5   Domain = $^-\infty,\infty$ Range = -4, 4

Domain = $^-\infty,\infty$ Range = $2,\infty$

Question #6   Domain = $^-\infty,\infty$ Range = 5

Question #7   (3,9), (-4,16), (6,3), (1,9), (1,3)

Competency 0005, *page 48*

Question #1 $x^2 - 10x + 25$

Question #2 $25x^2 - 10x - 48$

Question #3 $x^2 - 9x - 36$

Competency 0006, *page 50*

Question #1 $S_5 = 75$

Question #2 $S_n = 28$

Question #3 $S_n = -\dfrac{-31122}{15625} \approx\ ^- 1.99$

page 53

Question #1   Question #2

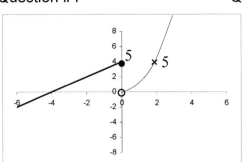   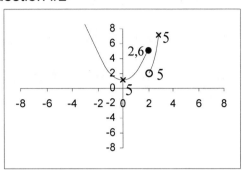

page 67

$$\cot\theta = \frac{x}{y}$$

Question #1

$$\frac{x}{y} = \frac{x}{r} \times \frac{r}{y} = \frac{x}{y} = \cot\theta$$

$$1 + \cot^2\theta = \csc^2\theta$$

Question #2

$$\frac{y^2}{y^2} + \frac{x^2}{y^2} = \frac{r^2}{y^2} = \csc^2\theta$$

Competency 0007, *page 79*

Question # 1   49.34

Question # 2   1

page 100

Question #1 $g(x) = -20\sin x + c$

Question #2 $g(x) = \pi\sec x + c$

Competency 0008, *page 115*

Question #1 It takes Curly 15 minutes to paint the elephant alone

Question #2 The original number is 5/15

Question #3 The car was traveling at 68mph and the truck was traveling
            at 62mph

Sample Test

Directions: Read each item and select the best response.

1. Change $.\overline{63}$ into a fraction in simplest form.
    (Easy) (Competency 0001)

    A) 63/100
    B) 7/11
    C) 6 3/10
    D) 2/3

2. Which of the following is always composite if *x* is odd, *y* is even, and both *x* and *y* are greater than or equal to 2?
    (Average Rigor) (Competency 0001)

    A) $x + y$
    B) $3x + 2y$
    C) $5xy$
    D) $5x + 3y$

3. Solve for x: $18 = 4 + |2x|$
    (Rigorous) (Competency 0001)

    A) $\{-11, 7\}$
    B) $\{-7, 0, 7\}$
    C) $\{-7, 7\}$
    D) $\{-11, 11\}$

4. Find the LCM of 27, 90 and 84.
    (Easy) (Competency 0001)

    A) 90
    B) 3780
    C) 204120
    D) 1260

5. Which of the following sets is closed under division?
    (Average Rigor) (Competency 0002)

    I) $\{½, 1, 2, 4\}$
    II) $\{-1, 1\}$
    III) $\{-1, 0, 1\}$

    A) I only
    B) II only
    C) III only
    D) I and II

6. Which of the following illustrates an inverse property?
    (Easy) (Competency 0002)

    A) a + b = a - b
    B) a + b = b + a
    C) a + 0 = a
    D) a + (-a) = 0

7. Evaluate $3^{1/2}(9^{1/3})$
    (Rigorous) (Competency 0002)

    A) $27^{5/6}$
    B) $9^{7/12}$
    C) $3^{5/6}$
    D) $3^{6/7}$

8. Simplify: $\dfrac{10}{1+3i}$

(Average Rigor) (Competency 0002)

A) $-1.25(1-3i)$
B) $1.25(1+3i)$
C) $1+3i$
D) $1-3i$

9. Which of the following is equivalent to $\sqrt[b]{x^a}$?

(Easy) (Competency 0002)

A) $x^{a/b}$
B) $x^{b/a}$
C) $a^{x/b}$
D) $b^{x/a}$

10. Which of the following is incorrect?

(Rigorous) (Competency 0002)

A) $(x^2y^3)^2 = x^4y^6$
B) $m^2(2n)^3 = 8m^2n^3$
C) $(m^3n^4)/(m^2n^2) = mn^2$
D) $(x+y^2)^2 = x^2 + y^4$

11. Simplify: $\sqrt{27} + \sqrt{75}$

(Rigorous) (Competency 0002)

A) $8\sqrt{3}$
B) 34
C) $34\sqrt{3}$
D) $15\sqrt{3}$

12. Determine the rectangular coordinates of the point with polar coordinates (5, 60°).

(Average Rigor) (Competency 0002)

A) (0.5, 0.87)
B) (-0.5, 0.87)
C) (2.5, 4.33)
D) (25, 150°)

13. Which of the following is a factor of the expression $9x^2 + 6x - 35$?

(Rigorous) (Competency 0003)

A) 3x-5
B) 3x-7
C) x+3
D) x-2

14. $f(x) = 3x - 2;\ f^{-1}(x) =$

(Average Rigor) (Competency 0003)

A) $3x + 2$
B) $x/6$
C) $2x - 3$
D) $(x+2)/3$

15. State the domain of the function $f(x) = \dfrac{3x-6}{x^2-25}$

(Average Rigor) (Competency 0003)

A) $x \neq 2$
B) $x \neq 5, -5$
C) $x \neq 2, -2$
D) $x \neq 5$

16. Which of the following is a factor of $6 + 48m^3$
(Rigorous) (Competency 0003)

A) $(1 + 2m)$
B) $(1 - 8m)$
C) $(1 + m - 2m)$
D) $(1 - m + 2m)$

17. Given $f(x) = 3x - 2$ and $g(x) = x^2$, determine $g(f(x))$.
(Average Rigor) (Competency 0003)

A) $3x^2 - 2$
B) $9x^2 + 4$
C) $9x^2 - 12x + 4$
D) $3x^3 - 2$

18. What is the slope of any line parallel to the line 2x + 4y = 4?
(Easy) (Competency 0003)

A) -2
B) -1
C) – ½
D) 2

19. Solve the system of equations for x, y and z.
(Rigorous) (Competency 0004)

$$3x + 2y - z = 0$$
$$2x + 5y = 8z$$
$$x + 3y + 2z = 7$$

A) $(-1, 2, 1)$
B) $(1, 2, -1)$
C) $(-3, 4, -1)$
D) $(0, 1, 2)$

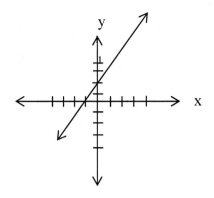

20. What is the equation of the above graph?
(Easy) (Competency 0004)

A) $2x + y = 2$
B) $2x - y = -2$
C) $2x - y = 2$
D) $2x + y = -2$

21. Given a vector with horizontal component 5 and vertical component 6, determine the length of the vector.
(Average Rigor) (Competency 0004)

A) 61
B) $\sqrt{61}$
C) 30
D) $\sqrt{30}$

22. What would be the shortest method of solution for the system of equations below? *(Easy) (Competency 0004)*

$$3x + 2y = 38$$
$$4x + 8 = y$$

A) linear combination
B) additive inverse
C) substitution
D) graphing

23. Solve for x by factoring $2x^2 - 3x - 2 = 0$. *(Average Rigor) (Competency 0005)*

A) x = (-1,2)
B) x = (0.5,-2)
C) x=(-0.5,2)
D) x=(1,-2)

24. Find the sum of the first one hundred terms in the progression.
(-6, -2, 2 . . .)
(Rigorous) (Competency 0005)

A) 19,200
B) 19,400
C) -604
D) 604

25. What is the sum of the first 20 terms of the geometric sequence (2,4,8,16,32,…)? *(Average Rigor) (Competency 0005)*

A) 2097150
B) 1048575
C) 524288
D) 1048576

26. Which graph represents the equation of $y = x^2 + 3x$? *(Average Rigor) (Competency 0005)*

A)   B)

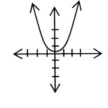 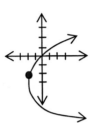

C)   D)

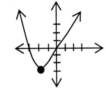 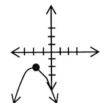

27. Which graph represents the solution set for $x^2 - 5x > -6$? *(Average Rigor) (Competency 0005)*

A)
    -2   0   2

B)
    -3   0   3

C)
    -2   0   2

D)
    -3   0 2 3

28. Which equation corresponds to the logarithmic statement: $\log_x k = m$?
(Rigorous) (Competency 0006)

A) $x^m = k$
B) $k^m = x$
C) $x^k = m$
D) $m^x = k$

29. How does the function $y = x^3 + x^2 + 4$ behave from $x = 1$ to $x = 3$?
(Average Rigor) (Competency 0006)

A) increasing, then decreasing
B) increasing
C) decreasing
D) neither increasing nor decreasing

30. Find the absolute maximum obtained by the function $y = 2x^2 + 3x$ on the interval $x = 0$ to $x = 3$.
(Rigorous) (Competency 0006)

A) $-3/4$
B) $-4/3$
C) 0
D) 27

31. Solve for x $10^{x-3} + 5 = 105$.
(Rigorous) (Competency 0006)

A) 3
B) 10
C) 2
D) 5

32. Which expression is equivalent to $1 - \sin^2 x$?
(Rigorous) (Competency 0007)

A) $1 - \cos^2 x$
B) $1 + \cos^2 x$
C) $1/\sec x$
D) $1/\sec^2 x$

33. Determine the measures of angles A and B.
(Average Rigor) (Competency 0007)

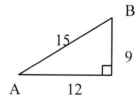

A) A = 30°, B = 60°
B) A = 60°, B = 30°
C) A = 53°, B = 37°
D) A = 37°, B = 53°

34. Which expression is not equal to sinx?
(Average Rigor) (Competency 0007)

A) $\sqrt{1 - \cos^2 x}$
B) $\tan x \cos x$
C) $1/\csc x$
D) $1/\sec x$

35. For an acute angle x, sinx = 3/5. What is cotx?
(Rigorous) (Competency 0007)

A) 5/3
B) 3/4
C) 1.33
D) 1

MATHEMATICS   196

36. **Find the following limit:**

$$\lim_{x \to 2} \frac{x^2 - 4}{x - 2}$$

(Average Rigor) (Competency 0008)

A) 0
B) Infinity
C) 2
D) 4

37. **If the area of the base of a cone is tripled, the volume will be**
(Rigorous) (Competency 0008)

A) the same as the original
B) 9 times the original
C) 3 times the original
D) 3 π times the original

38. **The mass of a Chips Ahoy cookie would be to**
(Average Rigor) (Competency 0008)

A) 1 kilogram
B) 1 gram
C) 15 grams
D) 15 milligrams

39. **Find the first derivative of the function:**

$$f(x) = x^3 - 6x^2 + 5x + 4$$

(Rigorous) (Competency 0008)

A) $3x^3 - 12x^2 + 5x = f'(x)$
B) $3x^2 - 12x - 5 = f'(x)$
C) $3x^2 - 12x + 9 = f'(x)$
D) $3x^2 - 12x + 5 = f'(x)$

40. **Differentiate:** $y = e^{3x+2}$
(Rigorous) (Competency 0008)

A) $3e^{3x+2} = y'$
B) $3e^{3x} = y'$
C) $6e^3 = y'$
D) $(3x + 2)e^{3x+1} = y'$

41. **Find the slope of the line tangent to** $y = 3x(\cos x)$ **at** $(\pi/2,\ \pi/2)$.
(Rigorous) (Competency 0008)

A) $-3\pi/2$
B) $3\pi/2$
C) $\pi/2$
D) $-\pi/2$

42. **Find the equation of the line tangent to** $y = 3x^2 - 5x$ **at** $(1, -2)$.
(Rigorous) (Competency 0008)

A) $y = x - 3$
B) $y = 1$
C) $y = x + 2$
D) $y = x$

43. **Find the antiderivative for** $4x^3 - 2x + 6 = y$.
(Rigorous) (Competency 0008)

A) $x^4 - x^2 + 6x + C$
B) $x^4 - 2/3x^3 + 6x + C$
C) $12x^2 - 2 + C$
D) $4/3x^4 - x^2 + 6x + C$

44. Find the antiderivative for the function $y = e^{3x}$.
(Rigorous) (Competency 0008)

A) $3x(e^{3x}) + C$
B) $3(e^{3x}) + C$
C) $1/3(e^{x}) + C$
D) $1/3(e^{3x}) + C$

45. The acceleration of a particle is dv/dt = 6 m/s². Find the velocity at t=10 given an initial velocity of 15 m/s.
(Average Rigor) (Competency 0008)

A) 60 m/s
B) 150 m/s
C) 75 m/s
D) 90 m/s

46. If the velocity of a body is given by v = 16 - t², find the distance traveled from t = 0 until the body comes to a complete stop.
(Average Rigor) (Competency 0008)

A) 16
B) 43
C) 48
D) 64

47. Evaluate $\int_0^2 (x^2 + x - 1) dx$
(Rigorous) (Competency 0008)

A) 11/3
B) 8/3
C) -8/3
D) -11/3

48. Evaluate: $\int (x^3 + 4x - 5) dx$
(Rigorous) (Competency 0008)

A) $3x^2 + 4 + C$
B) $\frac{1}{4}x^4 - 2/3x^3 + 6x + C$
C) $x^{4/3} + 4x - 5x + C$
D) $x^3 + 4x^2 - 5x + C$

49. Find the area under the function $y = x^2 + 4$ from $x = 3$ to $x = 6$.
(Average Rigor) (Competency 0008)

A) 75
B) 21
C) 96
D) 57

50. Compute the area of the shaded region, given a radius of 5 meters. 0 is the center.
(Rigorous) (Competency 0009)

A) 7.13 cm²
B) 7.13 m²
C) 78.5 m²
D) 19.63 m²

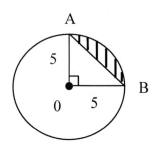

51. Find the area of the figure pictured below.
(Rigorous) (Competency 0009)

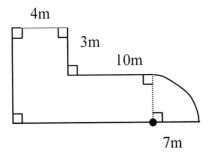

4m
3m
10m
7m

A) 136.47 m²
B) 148.48 m²
C) 293.86 m²
D) 178.47 m²

52. Find the surface area of a box which is 3 feet wide, 5 feet tall, and 4 feet deep.
(Easy) (Competency 0009)

A) 47 sq. ft.
B) 60 sq. ft.
C) 94 sq. ft
D) 188 sq. ft.

53. Given a 30 meter x 60 meter garden with a circular fountain with a 5 meter radius, calculate the area of the portion of the garden not occupied by the fountain.
(Average Rigor) (Competency 0009)

A) 1721 m²
B) 1879 m²
C) 2585 m²
D) 1015 m²

54. Find the height of a box with surface area of 94 sq. ft. with a width of 3 feet and a depth of 4 feet.
(Average Rigor) (Competency 0009)

A) 3 ft.
B) 4 ft.
C) 5 ft
D) 6 ft.

55. If a ship sails due south 6 miles, then due west 8 miles, how far was it from the starting point?
(Average Rigor) (Competency 0010)

A) 100 miles
B) 10 miles
C) 14 miles
D) 48 miles

56. What is the measure of minor arc AD, given measure of arc PS is 40° and $m < K = 10°$?
(Rigorous) (Competency 0010)

A) 50°
B) 20°
C) 30°
D) 25°

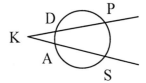

57. Which theorem can be used to prove $\triangle BAK \cong \triangle MKA$?
(Average Rigor) (Competency 0010)

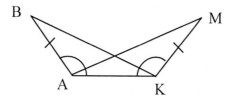

   A) SSS
   B) ASA
   C) SAS
   D) AAS

58. Determine the area of the shaded region of the trapezoid in terms of *x* and *y*.
(Rigorous) (Competency 0010)

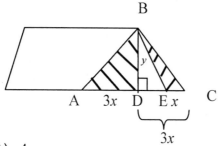

   A) $4xy$
   B) $2xy$
   C) $3x^2 y$
   D) There is not enough information given.

59. What is the sum of the interior angles of the following triangle?
(Easy) (Competency 0010)

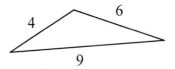

   A) 90°
   B) 180°
   C) 270°
   D) 360°

60. Ginny and Nick head back to their respective colleges after being home for the weekend. They leave their house at the same time and drive for 4 hours. Ginny drives due south at the average rate of 60 miles per hour and Nick drives due east at the average rate of 60 miles per hour. What is the straight-line distance between them, in miles, at the end of the 4 hours?
(Average Rigor) (Competency 0010)

   A) $120\sqrt{2}$
   B) 240
   C) $240\sqrt{2}$
   D) 288

61. Given that QO⊥NP and QO=NP, quadrilateral NOPQ can most accurately be described as a
(Easy) (Competency 0011)

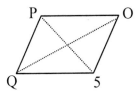

A) parallelogram
B) rectangle
C) square
D) rhombus

62. Given $K(-4, y)$ and $M(2, -3)$ with midpoint $L(x, 1)$, determine the values of x and y.
(Rigorous) (Competency 0011)

A) $x = -1, y = 5$
B) $x = 3, y = 2$
C) $x = 5, y = -1$
D) $x = -1, y = -1$

63. Find the length of the major axis of $x^2 + 9y^2 = 36$.
(Rigorous) (Competency 0011)

A) 4
B) 6
C) 12
D) 8

64. Which equation represents a circle with a diameter whose endpoints are $(0, 7)$ and $(0, 3)$?
(Rigorous) (Competency 0011)

A) $x^2 + y^2 + 21 = 0$
B) $x^2 + y^2 - 10y + 21 = 0$
C) $x^2 + y^2 - 10y + 9 = 0$
D) $x^2 - y^2 - 10y + 9 = 0$

65. Compute the median for the following data set:
(Easy) (Competency 0012)

{12, 19, 13, 16, 17, 14}

A) 14.5
B) 15.17
C) 15
D) 16

66. Half the students in a class scored 80% on an exam, most of the rest scored 85% except for one student who scored 10%. Which would be the best measure of central tendency for the test scores?
(Rigorous) (Competency 0012)

A) mean
B) median
C) mode
D) either the median or the mode because they are equal

67. What conclusion can be drawn from the graph below?

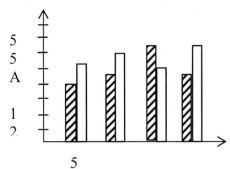

MLK Elementary Student Enrollment   ▨ **Girls**   ☐ **Boys**
(Easy) (Competency 0012)

A) The number of students in first grade exceeds the number in second grade.

B) There are more boys than girls in the entire school.

C) There are more girls than boys in the first grade.

D) Third grade has the largest number of students.

68. The pie chart below shows sales at an automobile dealership for the first four months of a year. What percentage of the vehicles were sold in April?
(Easy) (Competency 0012)

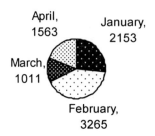

A) More than 50%
B) Less than 25%
C) Between 25% and 50%
D) None

69. How many ways are there to choose a potato and two green vegetables from a choice of three potatoes and seven green vegetables?
(Average Rigor) (Competency 0013)

A) 126
B) 63
C) 21
D) 252

70. Determine the number of subsets of set *K*.
K = {4, 5, 6, 7}
(Average Rigor) (Competency 0013)

A) 15
B) 16
C) 17
D) 18

71. A jar contains 3 red marbles, 5 white marbles, 1 green marble and 15 blue marbles. If one marble is picked at random from the jar, what is the probability that it will be red?
(Easy) (Competency 0013)

A) 1/3
B) 1/8
C) 3/8
D) 1/24

72. If there are three people in a room, what is the probability that at least two of them will share a birthday? (Assume a year has 365 days)
(Rigorous) (Competency 0013)

A) 0.67
B) 0.05
C) 0.008
D) 0.33

73. Compute the standard deviation for the following set of temperatures.
(37, 38, 35, 37, 38, 40, 36, 39)
(Easy) (Competency 0014)

A) 37.5
B) 1.5
C) 0.5
D) 2.5

74. Given the series of examples below, what is 5¢4?
(Average Rigor) (Competency 0015)

4¢3=13   7¢2=47
3¢1=8   1¢5=-4

A) 20
B) 29
C) 1
D) 21

75. What would be the total cost of a suit for $295.99 and a pair of shoes for $69.95 including 6.5% sales tax?
(Average Rigor) (Competency 0015)

A) $389.73
B) $398.37
C) $237.86
D) $315.23

76. A student had 60 days to appeal the results of an exam. If the results were received on March 23, what was the last day that the student could appeal?
(Average Rigor) (Competency 0015)

A) May 21
B) May 22
C) May 23
D) May 24

77. The volume of water flowing through a pipe varies directly with the square of the radius of the pipe. If the water flows at a rate of 80 liters per minute through a pipe with a radius of 4 cm, at what rate would water flow through a pipe with a radius of 3 cm?
(Rigorous) (Competency 0015)

A) 45 liters per minute
B) 6.67 liters per minute
C) 60 liters per minute
D) 4.5 liters per minute

78. If three cups of concentrate are needed to make 2 gallons of fruit punch, how many cups are needed to make 5 gallons?
(Easy) (Competency 0015)

A) 6 cups
B) 7 cups
C) 7.5 cups
D) 10 cups

79. Which of the following best describes the process of induction?
(Average Rigor) (Competency 0016)

A) making an inference based on a set of universal laws
B) making an inference based on conjecture
C) making an inference based on a set of concrete examples
D) making an inference based on mathematical principles

80. Which of the following is a valid argument?
(Average Rigor) (Competency 0016)

A) Given: if p then q; q. Therefore: p.
B) Given: if p then q; ~p. Therefore: ~q.
C) Given: if p then q; p. Therefore: q.
D) Given: if p then q; ~q. Therefore: p.

Answer Key

1)	B	17)	C	33)	D	49)	A	65)	C
2)	C	18)	C	34)	D	50)	B	66)	B
3)	C	19)	A	35)	B	51)	B	67)	B
4)	B	20)	B	36)	D	52)	C	68)	B
5)	B	21)	B	37)	C	53)	A	69)	A
6)	D	22)	C	38)	C	54)	C	70)	B
7)	B	23)	C	39)	D	55)	B	71)	B
8)	D	24)	A	40)	A	56)	B	72)	C
9)	A	25)	A	41)	A	57)	C	73)	B
10)	D	26)	C	42)	A	58)	B	74)	D
11)	A	27)	D	43)	A	59)	B	75)	A
12)	C	28)	A	44)	D	60)	C	76)	B
13)	A	29)	B	45)	C	61)	C	77)	A
14)	D	30)	D	46)	B	62)	A	78)	C
15)	B	31)	D	47)	B	63)	C	79)	C
16)	A	32)	D	48)	B	64)	B	80)	C

Rigor Table

	Easy %20	Average Rigor %40	Rigorous %40
Question #	1, 4, 6, 9, 18, 20, 22, 52, 59, 61, 65, 67, 68, 71, 73, 78	2, 5, 8, 12, 14, 15, 17, 21, 23, 25, 26, 27, 29, 33, 34, 36, 38, 45, 46, 49, 53, 54, 55, 57, 60, 69, 70, 74, 75, 76, 79, 80	3, 7, 10, 11, 13, 16, 19, 24, 28, 30, 31, 32, 35, 37, 39, 40, 41, 42, 43, 44, 47, 48, 50, 51, 56, 58, 62, 63, 64, 66, 72, 77

Rationales with Sample Questions

1. Change $.\overline{63}$ into a fraction in simplest form.
    (Easy) (Competency 0001)

A) 63/100
B) 7/11
C) 6 3/10
D) 2/3

Answer: B

Let N = .636363…. Then multiplying both sides of the equation by 100 or 10^2 (because there are 2 repeated numbers), we get 100N = 63.636363… Then subtracting the two equations gives 99N = 63 or N = $\frac{63}{99} = \frac{7}{11}$.

2. Which of the following is always composite if *x* is odd, *y* is even, and both *x* and *y* are greater than or equal to 2?
    (Average Rigor) (Competency 0001)

A) $x+y$
B) $3x+2y$
C) $5xy$
D) $5x+3y$

Answer: C

A composite number is a number which is not prime. The prime number sequence begins 2,3,5,7,11,13,17,…. To determine which of the expressions is <u>always</u> composite, experiment with different values of x and y, such as x=3 and y=2, or x=5 and y=2. It turns out that 5xy will always be an even number, and therefore, composite, if y=2.

3. **Solve for** x: $18 = 4 + |2x|$
   (Rigorous) (Competency 0001)

A) $\{-11, 7\}$

B) $\{-7, 0, 7\}$

C) $\{-7, 7\}$

D) $\{-11, 11\}$

Answer: C

Using the definition of absolute value, two equations are possible: 18 = 4 + 2x or 18 = 4 – 2x. Solving for x gives x = 7 or x = -7.

4. **Find the LCM of 27, 90 and 84.**
   (Easy) (Competency 0001)

A) 90

B) 3780

C) 204120

D) 1260

Answer: B

To find the LCM of the above numbers, factor each into its prime factors and multiply each common factor the maximum number of times it occurs. Thus 27=3x3x3; 90=2x3x3x5; 84=2x2x3x7; LCM = 2x2x3x3x3x5x7=3780.

5. **Which of the following sets is closed under division?**
   (Average Rigor) (Competency 0002)

   I) {½, 1, 2, 4}
   II) {-1, 1}
   III) {-1, 0, 1}

A) I only
B) II only
C) III only
D) I and II

Answer: B

I is not closed because $\dfrac{4}{.5} = 8$ and 8 is not in the set.

III is not closed because $\dfrac{1}{0}$ is undefined.

II is closed because $\dfrac{-1}{1} = -1, \dfrac{1}{-1} = -1, \dfrac{1}{1} = 1, \dfrac{-1}{-1} = 1$ and all the answers are in the set.

6. **Which of the following illustrates an inverse property?**
   (Easy) (Competency 0002)

A) a + b = a - b
B) a + b = b + a
C) a + 0 = a
D) a + (-a) = 0

Answer: D

Because a + (-a) = 0 is a statement of the Additive Inverse Property of Algebra.

7. **Evaluate** $3^{1/2}(9^{1/3})$
   (Rigorous) (Competency 0002)

A) $27^{5/6}$
B) $9^{7/12}$
C) $3^{5/6}$
D) $3^{6/7}$

Answer: B

Getting the bases the same gives us $3^{\frac{1}{2}}3^{\frac{2}{3}}$. Adding exponents gives $3^{\frac{7}{6}}$. Then some additional manipulation of exponents produces $3^{\frac{7}{6}} = 3^{\frac{14}{12}} = \left(3^2\right)^{\frac{7}{12}} = 9^{\frac{7}{12}}$.

8. **Simplify:** $\dfrac{10}{1+3i}$
   (Average Rigor) (Competency 0002)

A) $-1.25(1-3i)$
B) $1.25(1+3i)$
C) $1+3i$
D) $1-3i$

Answer: D

Multiplying numerator and denominator by the conjugate gives
$$\frac{10}{1+3i} \times \frac{1-3i}{1-3i} = \frac{10(1-3i)}{1-9i^2} = \frac{10(1-3i)}{1-9(-1)} = \frac{10(1-3i)}{10} = 1-3i.$$

9. **Which of the following is equivalent to $\sqrt[b]{x^a}$?**
   (Easy) (Competency 0002)

A) $x^{a/b}$
B) $x^{b/a}$
C) $a^{x/b}$
D) $b^{x/a}$

Answer: A

The bth root, expressed in the form $\sqrt[b]{}$, can also be written as an exponential, $1/b$. Writing the expression in this form, $\left(x^a\right)^{1/b}$, and then multiplying exponents yields $x^{a/b}$.

10. **Which of the following is incorrect?**
    (Rigorous) (Competency 0002)

A) $(x^2 y^3)^2 = x^4 y^6$
B) $m^2 (2n)^3 = 8m^2 n^3$
C) $(m^3 n^4)/(m^2 n^2) = mn^2$
D) $(x + y^2)^2 = x^2 + y^4$

Answer: D

Using FOIL to do the expansion, we get $(x + y^2)^2 = (x + y^2)(x + y^2) = x^2 + 2xy^2 + y^4$.

11. **Simplify:** $\sqrt{27} + \sqrt{75}$
    (Rigorous) (Competency 0002)

A) $8\sqrt{3}$
B) 34
C) $34\sqrt{3}$
D) $15\sqrt{3}$

Answer: A

Simplifying radicals gives $\sqrt{27} + \sqrt{75} = 3\sqrt{3} + 5\sqrt{3} = 8\sqrt{3}$.

12. Determine the rectangular coordinates of the point with polar coordinates (5, 60°).
    (Average Rigor) (Competency 0002)

A) (0.5, 0.87)
B) (-0.5, 0.87)
C) (2.5, 4.33)
D) (25, 150°)

Answer: C

Given the polar point $(r, \theta) = (5, 60)$, we can find the rectangular coordinates this way: $(x,y) = (r\cos\theta, r\sin\theta) = (5\cos 60, 5\sin 60) = (2.5, 4.33)$.

13. Which of the following is a factor of the expression $9x^2 + 6x - 35$?
    (Rigorous) (Competency 0003)

A) 3x-5
B) 3x-7
C) x+3
D) x-2

Answer: A

Recognize that the given expression can be written as the sum of two squares and utilize the formula $a^2 - b^2 = (a+b)(a-b)$.

$9x^2 + 6x - 35 = (3x + 1)^2 - 36 = (3x + 1 + 6)(3x + 1 - 6) = (3x + 7)(3x - 5)$

14. $f(x) = 3x - 2;\ f^{-1}(x) =$
    (Average Rigor) (Competency 0003)

A) $3x + 2$
B) $x / 6$
C) $2x - 3$
D) $(x + 2)/3$

Answer: D

To find the inverse, f⁻¹(x), of the given function, reverse the variables in the given equation, y = 3x − 2, to get x = 3y − 2. Then solve for y as follows:

x+2 = 3y, and y = $\dfrac{x+2}{3}$.

15. State the domain of the function $f(x) = \dfrac{3x-6}{x^2-25}$
    (Average Rigor) (Competency 0003)

A) $x \neq 2$
B) $x \neq 5, -5$
C) $x \neq 2, -2$
D) $x \neq 5$

Answer: B

The values of 5 and −5 must be omitted from the domain of all real numbers because if x took on either of those values, the denominator of the fraction would have a value of 0, and therefore the fraction would be undefined.

16. Which of the following is a factor of $6 + 48m^3$
    (Rigorous) (Competency 0003)

A) (1 + 2m)
B) (1 - 8m)
C) (1 + m - 2m)
D) (1 - m + 2m)

Answer: A

Removing the common factor of 6 and then factoring the sum of two cubes gives
$6 + 48m^3 = 6(1 + 8m^3) = 6(1 + 2m)(1^2 - 2m + (2m)^2)$.

17. Given $f(x) = 3x - 2$ **and** $g(x) = x^2$, **determine** $g(f(x))$.
    (Average Rigor) (Competency 0003)

A) $3x^2 - 2$
B) $9x^2 + 4$
C) $9x^2 - 12x + 4$
D) $3x^3 - 2$

Answer: C

The composite function g(f(x)) = $(3x-2)^2$ = $9x^2 - 12x + 4$.

18. **What is the slope of any line parallel to the line 2x + 4y = 4?**
(Easy) (Competency 0003)

A) -2
B) -1
C) – ½
D) 2

Answer: C

The formula for slope is y = mx + b, where m is the slope. Lines that are parallel have the same slope.

$$2x + 4y = 4$$
$$4y = -2x + 4$$
$$y = \frac{-2x}{4} + 1$$
$$y = \frac{-1}{2}x + 1$$

19. **Solve the system of equations** for x, y **and** z.
(Rigorous) (Competency 0004)

$$3x + 2y - z = 0$$
$$2x + 5y = 8z$$
$$x + 3y + 2z = 7$$

A) $(-1, \ 2, \ 1)$
B) $(1, \ 2, \ -1)$
C) $(-3, \ 4, \ -1)$
D) $(0, \ 1, \ 2)$

Answer: A

Multiplying equation 1 by 2, and equation 2 by –3, and then adding together the two resulting equations gives -11y + 22z = 0. Solving for y gives y = 2z. In the meantime, multiplying equation 3 by –2 and adding it to equation 2 gives –y – 12z = -14. Then substituting 2z for y, yields the result z = 1. Subsequently, one can easily find that y = 2, and x = -1.

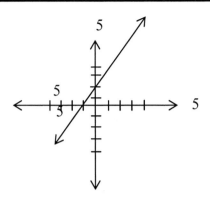

20. What is the equation of the above graph?
(Easy) (Competency 0004)

A) $2x + y = 2$
B) $2x - y = -2$
C) $2x - y = 2$
D) $2x + y = -2$

Answer: B

By observation, we see that the graph has a y-intercept of 2 and a slope of 2/1 = 2. Therefore its equation is y = mx + b = 2x + 2. Rearranging the terms gives 2x − y = -2.

21. Given a vector with horizontal component 5 and vertical component 6, determine the length of the vector.
(Average Rigor) (Competency 0004)

A) 61
B) $\sqrt{61}$
C) 30
D) $\sqrt{30}$

Answer: B

Using the Pythagorean Theorem, we get v = $\sqrt{36 + 25} = \sqrt{61}$.

22. **What would be the shortest method of solution for the system of equations below?**
    (Easy) (Competency 0004)

    $$3x + 2y = 38$$
    $$4x + 8 = y$$

A) linear combination
B) additive inverse
C) substitution
D) graphing

Answer: C

Since the second equation is already solved for y, it would be easiest to use the substitution method.

23. **Solve for x by factoring** $2x^2 - 3x - 2 = 0$.
    (Average Rigor) (Competency 0005)

A) x = (-1,2)
B) x = (0.5,-2)
C) x=(-0.5,2)
D) x=(1,-2)

Answer: C

$2x^2 - 3x - 2 = 2x^2 - 4x + x - 2 = 2x(x - 2) + (x - 2) = (2x + 1)(x - 2) = 0.$
Thus x = -0.5 or 2.

24. **Find the sum of the first one hundred terms in the progression.**
    (-6, -2, 2 . . .)
    (Rigorous) (Competency 0005)

A) 19,200
B) 19,400
C) -604
D) 604

Answer: A

To find the 100th term: t_{100} = -6 + 99(4) = 390. To find the sum of the first 100 terms: S = $\frac{100}{2}(-6 + 390) = 19200$.

25. What is the sum of the first 20 terms of the geometric sequence (2,4,8,16,32,…)?
(Average Rigor) (Competency 0005)

A) 2097150
B) 1048575
C) 524288
D) 1048576

Answer: A

For a geometric sequence $a, ar, ar^2, ..., ar^n$, the sum of the first n terms is given by $\dfrac{a(r^n - 1)}{r - 1}$. In this case a=2 and r=2. Thus the sum of the first 20 terms of the sequence is given by $\dfrac{2(2^{20} - 1)}{2 - 1} = 2097150$.

26. **Which graph represents the equation of** $y = x^2 + 3x$ **?**
    (Average Rigor) (Competency 0005)

A)   B)

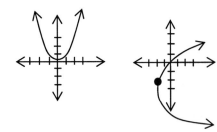

C)   D)

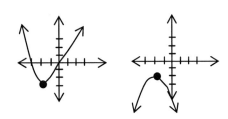

Answer: C

B is not the graph of a function. D is the graph of a parabola where the coefficient of x^2 is negative. A appears to be the graph of $y = x^2$. To find the x-intercepts of $y = x^2 + 3x$, set $y = 0$ and solve for x: $0 = x^2 + 3x = x(x + 3)$ to get x = 0 or x = -3. Therefore, the graph of the function intersects the x-axis at x=0 and x=-3.

27. Which graph represents the solution set for $x^2 - 5x > -6$?
    (Average Rigor) (Competency 0005)

A)

     5   5

B)

     5   5

C)

     5   5

D)

     5   5

Answer: D

Rewriting the inequality gives $x^2 - 5x + 6 > 0$. Factoring gives $(x - 2)(x - 3) > 0$. The two cut-off points on the number line are now at $x = 2$ and $x = 3$. Choosing a random number in each of the three parts of the numberline, we test them to see if they produce a true statement. If $x = 0$ or $x = 4$, $(x-2)(x-3)>0$ is true. If $x = 2.5$, $(x-2)(x-3)>0$ is false. Therefore the solution set is all numbers smaller than 2 or greater than 3.

28. Which equation corresponds to the logarithmic statement:
    $\log_x k = m$?
    (Rigorous) (Competency 0006)

A) $x^m = k$
B) $k^m = x$
C) $x^k = m$
D) $m^x = k$

Answer: A

By definition of log form and exponential form, $\log_x k = m$ corresponds to $x^m = k$.

29. How does the function $y = x^3 + x^2 + 4$ **behave from** $x = 1$ **to** $x = 3$?
(Average Rigor) (Competency 0006)

A) increasing, then decreasing
B) increasing
C) decreasing
D) neither increasing nor decreasing

Answer: B

To find critical points, take the derivative, set it equal to 0, and solve for x.
f'(x) = 3x² + 2x = x(3x+2)=0. CP at x=0 and x=-2/3. Neither of these CP is on the interval from x=1 to x=3. Testing the endpoints: at x=1, y=6 and at x=3, y=38. Since the derivative is positive for all values of x from x=1 to x=3, the curve is increasing on the entire interval.

30. Find the absolute maximum obtained by the function $y = 2x^2 + 3x$ **on the interval** $x = 0$ **to** $x = 3$.
(Rigorous) (Competency 0006)

A) $-3/4$
B) $-4/3$
C) 0
D) 27

Answer: D

Find CP at x=-.75 as done in #29. Since the CP is not in the interval from x=0 to x=3, just find the values of the functions at the endpoints. When x=0, y=0, and when x=3, y = 27. Therefore 27 is the absolute maximum on the given interval.

31. Solve for x $10^{x-3} + 5 = 105$.
(Rigorous) (Competency 0006)

A) 3
B) 10
C) 2
D) 5

Answer: D

$10^{x-3} = 100$. Taking the logarithm to base 10 of both sides
$(x - 3)\log_{10} 10 = \log_{10} 100; x - 3 = 2; x = 5.$

32. Which expression is equivalent to $1 - \sin^2 x$?
   (Rigorous) (Competency 0007)

A) $1 - \cos^2 x$
B) $1 + \cos^2 x$
C) $1/\sec x$
D) $1/\sec^2 x$

Answer: D

Using the Pythagorean Identity, we know $\sin^2 x + \cos^2 x = 1$. Thus $1 - \sin^2 x = \cos^2 x$, which by definition is equal to $1/\sec^2 x$.

33. Determine the measures of angles A and B.
   (Average Rigor) (Competency 0007)

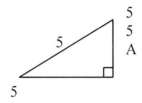

A) A = 30°, B = 60°
B) A = 60°, B = 30°
C) A = 53°, B = 37°
D) A = 37°, B = 53°

Answer: D

Tan A = 9/12=.75 and tan^{-1}.75 = 37 degrees. Since angle B is complementary to angle A, the measure of angle B is therefore 53 degrees.

34. **Which expression is not equal to sinx?**
    (Average Rigor) (Competency 0007)

A) $\sqrt{1 - \cos^2 x}$
B) $\tan x \cos x$
C) $1 / \csc x$
D) $1 / \sec x$

Answer: D

Using the basic definitions of the trigonometric functions and the Pythagorean identity, we see that the first three options are all identical to sinx. secx= 1/cosx is not the same as sinx.

35. **For an acute angle x, sinx = 3/5. What is cotx?**
    (Rigorous) (Competency 0007)

A) 5/3
B) 3/4
C) 1.33
D) 1

Answer: B

Using the Pythagorean Identity, we know $\sin^2 x + \cos^2 x = 1$. Thus

$$\cos x = \sqrt{1 - \frac{9}{25}} = \frac{4}{5}; \cot x = \frac{\cos x}{\sin x} = \frac{4}{3}.$$

36. **Find the following limit:** $\lim\limits_{x \to 2} \dfrac{x^2 - 4}{x - 2}$
    (Average Rigor) (Competency 0008)

A) 0
B) Infinity
C) 2
D) 4

Answer: D

First factor the numerator and cancel the common factor to get the limit.

$$\lim_{x \to 2} \frac{x^2 - 4}{x - 2} = \lim_{x \to 2} \frac{(x - 2)(x + 2)}{(x - 2)} = \lim_{x \to 2}(x + 2) = 4$$

37. If the area of the base of a cone is tripled, the volume will be
   (Rigorous) (Competency 0008)

A) the same as the original
B) 9 times the original
C) 3 times the original
D) 3π times the original

Answer: C

The formula for the volume of a cone is V = $\frac{1}{3}Bh$, where B is the area of the circular base and h is the height. If the area of the base is tripled, the volume becomes

V = $\frac{1}{3}(3B)h = Bh$, or three times the original area.

38. The mass of a Chips Ahoy cookie would be to
   (Average Rigor) (Competency 0008)

A) 1 kilogram
B) 1 gram
C) 15 grams
D) 15 milligrams

Answer: C

Since an ordinary cookie would not weigh as much as 1 kilogram, or as little as 1 gram or 15 milligrams, the only reasonable answer is 15 grams.

39. Find the first derivative of the function: $f(x) = x^3 - 6x^2 + 5x + 4$
   (Rigorous) (Competency 0008)

A) $3x^3 - 12x^2 + 5x = f'(x)$
B) $3x^2 - 12x - 5 = f'(x)$
C) $3x^2 - 12x + 9 = f'(x)$
D) $3x^2 - 12x + 5 = f'(x)$

Answer: D

Use the Power Rule for polynomial differentiation: if y = ax^n, then y'=nax^{n-1}.

40. **Differentiate:** $y = e^{3x+2}$
    (Rigorous) (Competency 0008)

A) $3e^{3x+2} = y'$
B) $3e^{3x} = y'$
C) $6e^3 = y'$
D) $(3x+2)e^{3x+1} = y'$

Answer: A

Use the Exponential Rule for derivatives of functions of e: if y = ae$^{f(x)}$, then
y' = f'(x)ae$^{f(x)}$. **Answer is A.**

41. **Find the slope of the line tangent to** $y = 3x(\cos x)$ **at** $(\pi/2, \pi/2)$.
    (Rigorous) (Competency 0008)

A) $-3\pi/2$
B) $3\pi/2$
C) $\pi/2$
D) $-\pi/2$

Answer: A

To find the slope of the tangent line, find the derivative, and then evaluate it at x

= $\dfrac{\pi}{2}$. y' = 3x(-sinx)+3cosx. At the given value of x,

y' = $3(\dfrac{\pi}{2})(-\sin\dfrac{\pi}{2}) + 3\cos\dfrac{\pi}{2} = \dfrac{-3\pi}{2}$.

42. **Find the equation of the line tangent to** $y = 3x^2 - 5x$ **at** $(1, -2)$.
    (Rigorous) (Competency 0008)

A) $y = x - 3$
B) $y = 1$
C) $y = x + 2$
D) $y = x$

Answer: A

To find the slope of the tangent line, find the derivative, and then evaluate it at x=1.

y'=6x-5=6(1)-5=1. Then using point-slope form of the equation of a line, we get y+2=1(x-1) or y = x-3.

43. **Find the antiderivative for** $4x^3 - 2x + 6 = y$.
    (Rigorous) (Competency 0008)

A) $x^4 - x^2 + 6x + C$
B) $x^4 - 2/3x^3 + 6x + C$
C) $12x^2 - 2 + C$
D) $4/3x^4 - x^2 + 6x + C$

Answer: A

Use the rule for polynomial integration: given ax^n, the antiderivative is $\dfrac{ax^{n+1}}{n+1}$.

44. **Find the antiderivative for the function** $y = e^{3x}$.
    (Rigorous) (Competency 0008)

A) $3x(e^{3x}) + C$
B) $3(e^{3x}) + C$
C) $1/3(e^x) + C$
D) $1/3(e^{3x}) + C$

Answer: D

Use the rule for integration of functions of e: $\int e^x dx = e^x + C$.

45. **The acceleration of a particle is dv/dt = 6 m/s². Find the velocity at t=10 given an initial velocity of 15 m/s.**
   (Average Rigor) (Competency 0008)

A) 60 m/s
B) 150 m/s
C) 75 m/s
D) 90 m/s

Answer: C

Recall that the derivative of the velocity function is the acceleration function. In reverse, the integral of the acceleration function is the velocity function. Therefore, if a=6, then v=6t+C. Given that at t=0, v=15, we get v = 6t+15. At t=10, v=60+15=75m/s.

46. **If the velocity of a body is given by v = 16 - t², find the distance traveled from t = 0 until the body comes to a complete stop.**
   (Average Rigor) (Competency 0008)

A) 16
B) 43
C) 48
D) 64

Answer: B

Recall that the derivative of the distance function is the velocity function. In reverse, the integral of the velocity function is the distance function. To find the time needed for the body to come to a stop when v=0, solve for t: $v = 16 - t^2 = 0$.

Result: t = 4 seconds. The distance function is s = 16t - $\dfrac{t^3}{3}$. At t=4, s= 64 − 64/3 or approximately 43 units.

47. Evaluate $\int_0^2 (x^2 + x - 1)dx$

(Rigorous) (Competency 0008)

A) 11/3
B) 8/3
C) -8/3
D) -11/3

Answer: B

Use the fundamental theorem of calculus to find the definite integral: given a continuous function f on an interval [a,b], then $\int_a^b f(x)dx = F(b) - F(a)$, where F is an antiderivative of f.

$\int_0^2 (x^2 + x - 1)dx = (\dfrac{x^3}{3} + \dfrac{x^2}{2} - x)$ Evaluate the expression at x=2, at x=0, and then subtract to get 8/3 + 4/2 – 2-0 = 8/3.

48. Evaluate: $\int (x^3 + 4x - 5)dx$

(Rigorous) (Competency 0008)

A) $3x^2 + 4 + C$
B) $\dfrac{1}{4}x^4 - 2/3x^3 + 6x + C$
C) $x^{4/3} + 4x - 5x + C$
D) $x^3 + 4x^2 - 5x + C$

Answer: B

Integrate as described in #43.

49. **Find the area under the function** $y = x^2 + 4$ **from** $x = 3$ **to** $x = 6$.
    (Average Rigor) (Competency 0008)

A) 75
B) 21
C) 96
D) 57

Answer: A

To find the area set up the definite integral: $\int\limits_{3}^{6}(x^2 + 4)dx = (\dfrac{x^3}{3} + 4x)$. Evaluate the expression at x=6, at x=3, and then subtract to get (72+24)-(9+12)=75.

50. **Compute the area of the shaded region, given a radius of 5 meters. 0 is the center.**
    (Rigorous) (Competency 0009)

A) 7.13 cm²
B) 7.13 m²
C) 78.5 m²
D) 19.63 m²

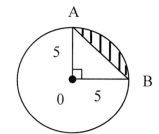

Answer: B

Area of triangle AOB is .5(5)(5) = 12.5 square meters. Since $\dfrac{90}{360} = .25$, the area of sector AOB (pie-shaped piece) is approximately .25(π)5² = 19.63. Subtracting the triangle area from the sector area to get the area of segment AB, we get approximately 19.63-12.5 = 7.13 square meters.

51. **Find the area of the figure pictured below.**
    (Rigorous) (Competency 0009)

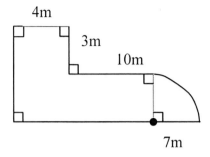

A) 136.47 m²
B) 148.48 m²
C) 293.86 m²
D) 178.47 m²

Answer: B

Divide the figure into 2 rectangles and one quarter circle. The tall rectangle on the left will have dimensions 10 by 4 and area 40. The rectangle in the center will have dimensions 7 by 10 and area 70. The quarter circle will have area $.25(\pi)7^2 = 38.48$.
The total area is therefore approximately 148.48.

52. **Find the surface area of a box which is 3 feet wide, 5 feet tall, and 4 feet deep.**
    (Easy) (Competency 0009)

A) 47 sq. ft.
B) 60 sq. ft.
C) 94 sq. ft
D) 188 sq. ft.

Answer: C

Let's assume the base of the rectangular solid (box) is 3 by 4, and the height is 5. Then the surface area of the top and bottom together is 2(12) = 24. The sum of the areas of the front and back are 2(15) = 30, while the sum of the areas of the sides are 2(20)=40. The total surface area is therefore 94 square feet.

53. Given a 30 meter x 60 meter garden with a circular fountain with a 5 meter radius, calculate the area of the portion of the garden not occupied by the fountain.
    (Average Rigor) (Competency 0009)

A) 1721 m²
B) 1879 m²
C) 2585 m²
D) 1015 m²

Answer: A

Find the area of the garden and then subtract the area of the fountain:
$30(60) - \pi(5)^2$ or approximately 1721 square meters.

54. Find the height of a box with surface area of 94 sq. ft. with a width of 3 feet and a depth of 4 feet.
    (Average Rigor) (Competency 0009)

A) 3 ft.
B) 4 ft.
C) 5 ft
D) 6 ft.

Answer: C

Set up the equation for the surface area in terms of the height of the box, and then solve for the height.

$94 = 2(3h) + 2(4h) + 2(12)$
$94 = 6h + 8h + 24$
$94 = 14h + 24$
$70 = 14h$
$5 = h$

55. **If a ship sails due south 6 miles, then due west 8 miles, how far was it from the starting point?**
    (Average Rigor) (Competency 0010)

A) 100 miles
B) 10 miles
C) 14 miles
D) 48 miles

Answer: B

Draw a right triangle with legs of 6 and 8. Find the hypotenuse using the Pythagorean Theorem. $6^2 + 8^2 = c^2$. Therefore, c = 10 miles.

56. **What is the measure of minor arc AD, given measure of arc PS is 40° and $m < K = 10°$?**
    (Rigorous) (Competency 0010)

A) 50°
B) 20°
C) 30°
D) 25°

Answer: B

The formula relating the measure of angle K and the two arcs it intercepts is $m\angle K = \frac{1}{2}(mPS - mAD)$. Substituting the known values, we get $10 = \frac{1}{2}(40 - mAD)$. Solving for mAD gives an answer of 20 degrees.

57. Which theorem can be used to prove $\triangle BAK \cong \triangle MKA$ **?**
(Average Rigor) (Competency 0010)

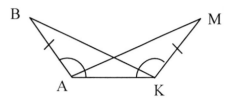

A) SSS
B) ASA
C) SAS
D) AAS

Answer: C

Since side AK is common to both triangles, the triangles can be proved congruent by using the Side-Angle-Side Postulate.

58. Determine the area of the shaded region of the trapezoid in terms of x and y.
(Rigorous) (Competency 0010)

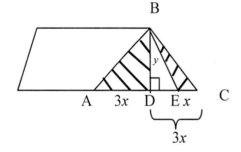

A) $4xy$

B) $2xy$

C) $3x^2y$

D) There is not enough information given.

Answer: B

To find the area of the shaded region, find the area of triangle ABC and then subtract the area of triangle DBE. The area of triangle ABC is .5(6x)(y) = 3xy. The area of triangle DBE is .5(2x)(y) = xy. The difference is 2xy.

59. **What is the sum of the interior angles of the following triangle?**
(Easy) (Competency 0010)

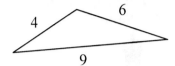

A) 90°
B) 180°
C) 270°
D) 360°

Answer: B

The sum of the interior angles of any Euclidean triangle is always 180°. Hence, the lengths of the sides of the triangle in the figure are irrelevant.

60. Ginny and Nick head back to their respective colleges after being home for the weekend. They leave their house at the same time and drive for 4 hours. Ginny drives due south at the average rate of 60 miles per hour and Nick drives due east at the average rate of 60 miles per hour. What is the straight-line distance between them, in miles, at the end of the 4 hours?
   (Average Rigor) (Competency 0010)

A) $120\sqrt{2}$
B) 240
C) $240\sqrt{2}$
D) 288

Answer: C

Draw a picture:

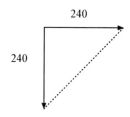

We have a right triangle, so we can use the Pythagorean Theorem to find the distance between the two points.

$$240^2 + 240^2 = c^2$$
$$2(240)^2 = c^2$$
$$240\sqrt{2} = c$$

61. **Given that QO⊥NP and QO=NP, quadrilateral NOPQ can most accurately be described as a**
   (Easy) (Competency 0011)

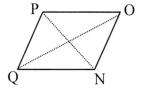

A) parallelogram
B) rectangle
C) square
D) rhombus

Answer: C

In an ordinary parallelogram, the diagonals are not perpendicular or equal in length. In a rectangle, the diagonals are not necessarily perpendicular. In a rhombus, the diagonals are not equal in length. In a square, the diagonals are both perpendicular and congruent.

62. **Given $K(-4, y)$ and $M(2, -3)$ with midpoint $L(x, 1)$, determine the values of x and y.**
   (Rigorous) (Competency 0011)

A) $x = -1, \; y = 5$
B) $x = 3, \; y = 2$
C) $x = 5, \; y = -1$
D) $x = -1, \; y = -1$

Answer: A

The formula for finding the midpoint (a,b) of a segment passing through the points $(x_1, y_1) \, and \, (x_2, y_2) \, is \, (a,b) = (\dfrac{x_1 + x_2}{2}, \dfrac{y_1 + y_2}{2})$. Setting up the corresponding equations from this information gives us $x = \dfrac{-4 + 2}{2}, \, and \, 1 = \dfrac{y - 3}{2}$. Solving for x and y gives x = -1 and y = 5.

63. **Find the length of the major axis of** $x^2 + 9y^2 = 36$.
    (Rigorous) (Competency 0011)

A) 4
B) 6
C) 12
D) 8

Answer: C

Dividing by 36, we get $\dfrac{x^2}{36} + \dfrac{y^2}{4} = 1$, which tells us that the ellipse intersects the x-axis at 6 and –6, and therefore the length of the major axis is 12. (The ellipse intersects the y-axis at 2 and –2).

64. **Which equation represents a circle with a diameter whose endpoints are** $(0,7)$ **and** $(0,3)$**?**
    (Rigorous) (Competency 0011)

A) $x^2 + y^2 + 21 = 0$
B) $x^2 + y^2 - 10y + 21 = 0$
C) $x^2 + y^2 - 10y + 9 = 0$
D) $x^2 - y^2 - 10y + 9 = 0$

Answer: B

With a diameter going from (0,7) to (0,3), the diameter of the circle must be 4, the radius must be 2, and the center of the circle must be at (0,5). Using the standard form for the equation of a circle, we get $(x-0)^2 + (y-5)^2 = 2^2$. Expanding, we get $x^2 + y^2 - 10y + 21 = 0$.

65. Compute the median for the following data set:
(Easy) (Competency 0012)

{12, 19, 13, 16, 17, 14}

A) 14.5
B) 15.17
C) 15
D) 16

Answer: C

Arrange the data in ascending order: 12,13,14,16,17,19. The median is the middle value in a list with an odd number of entries. When there is an even number of entries, the median is the mean of the two center entries. Here the average of 14 and 16 is 15.

66. Half the students in a class scored 80% on an exam, most of the rest scored 85% except for one student who scored 10%. Which would be the best measure of central tendency for the test scores?
(Rigorous) (Competency 0012)

A) mean
B) median
C) mode
D) either the median or the   mode because they are equal

Answer: B

In this set of data, the median (see #65) would be the most representative measure of central tendency since the median is independent of extreme values. Because of the 10% outlier, the mean (average) would be disproportionately skewed. In this data set, it is true that the median and the mode (number which occurs most often) are the same, but the median remains the best choice because of its special properties.

67. What conclusion can be drawn from the graph below?

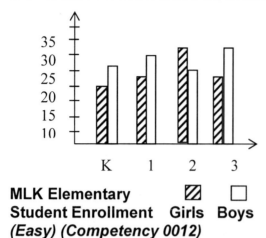

MLK Elementary
Student Enrollment   Girls   Boys
(Easy) (Competency 0012)

A) The number of students in first grade exceeds the number in second grade.
B) There are more boys than girls in the entire school.
C) There are more girls than boys in the first grade.
D) Third grade has the largest number of students.

Answer: B

In Kindergarten, first grade, and third grade, there are more boys than girls. The number of extra girls in grade two is more than made up for by the extra boys in all the other grades put together.

68. The pie chart below shows sales at an automobile dealership for the first four months of a year. What percentage of the vehicles were sold in April?
   (Easy) (Competency 0012)

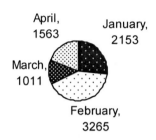

A) More than 50%
B) Less than 25%
C) Between 25% and 50%
D) None

Answer: B

It is clear from the chart that the April segment covers less than a quarter of the pie.

69. How many ways are there to choose a potato and two green vegetables from a choice of three potatoes and seven green vegetables?
   (Average Rigor) (Competency 0013)

A) 126
B) 63
C) 21
D) 252

Answer: A

There are 3 slots to fill. There are 3 choices for the first, 7 for the second, and 6 for the third. Therefore, the total number of choices is 3(7)(6) = 126.

70. **Determine the number of subsets of set *K*.**
    $K = \{4, 5, 6, 7\}$
    (Average Rigor) (Competency 0013)

A) 15
B) 16
C) 17
D) 18

Answer: B

A set of n objects has 2^n subsets. Therefore, here we have $2^4 = 16$ subsets. These subsets include four which each have 1 element only, six which each have 2 elements, four which each have 3 elements, plus the original set, and the empty set.

71. **A jar contains 3 red marbles, 5 white marbles, 1 green marble and 15 blue marbles. If one marble is picked at random from the jar, what is the probability that it will be red?**
    (Easy) (Competency 0013)

A) 1/3
B) 1/8
C) 3/8
D) 1/24

Answer: B

The total number of marbles is 24 and the number of red marbles is 3. Thus the probability of picking a red marble from the jar is 3/24=1/8.

72. **If there are three people in a room, what is the probability that at least two of them will share a birthday? (Assume a year has 365 days)**
    (Rigorous) (Competency 0013)

A) 0.67
B) 0.05
C) 0.008
D) 0.33

Answer: C

The best way to approach this problem is to use the fact that
the probability of an event + the probability of the event not happening = 1.
First find the probability that no two people will share a birthday and then subtract that from one.
The probability that two of the people will not share a birthday = 364/365 (since the second person's birthday can be one of the 364 days other than the birthday of the first person).
The probability that the third person will also not share either of the first two birthdays = (364/365) * (363/365) = 0.992.
Therefore, the probability that at least two people will share a birthday = 1 – 0.992= 0.008.

73. **Compute the standard deviation for the following set of temperatures. (37, 38, 35, 37, 38, 40, 36, 39)**
    (Easy) (Competency 0014)

A) 37.5
B) 1.5
C) 0.5
D) 2.5

Answer: B

Find the mean: 300/8 = 37.5. Then, using the formula for standard deviation, we get

$$\sqrt{\frac{2(37.5-37)^2 + 2(37.5-38)^2 + (37.5-35)^2 + (37.5-40)^2 + (37.5-36)^2 + (37.5-39)^2}{8}}$$

which has a value of 1.5.

74. Given the series of examples below, what is 5¢4?
    (Average Rigor) (Competency 0015)

$$4 \not\subset 3 = 13 \qquad 7 \not\subset 2 = 47$$
$$3 \not\subset 1 = 8 \qquad 1 \not\subset 5 = -4$$

A) 20
B) 29
C) 1
D) 21

Answer: D

By observation of the examples given, $a \not\subset b = a^2 - b$. Therefore, $5 \not\subset 4 = 25 - 4 = 21$.

75. What would be the total cost of a suit for $295.99 and a pair of shoes for $69.95 including 6.5% sales tax?
    (Average Rigor) (Competency 0015)

A) $389.73
B) $398.37
C) $237.86
D) $315.23

Answer: A

Before the tax, the total comes to $365.94. Then .065(365.94) = 23.79. With the tax added on, the total bill is 365.94 + 23.79 = $389.73. (Quicker way: 1.065(365.94) = 389.73.)

76. **A student had 60 days to appeal the results of an exam. If the results were received on March 23, what was the last day that the student could appeal?**
   (Average Rigor) (Competency 0015)

A) May 21
B) May 22
C) May 23
D) May 24

Answer: B

Recall: 30 days in April and 31 in March. 8 days in March + 30 days in April + 22 days in May brings him to a total of 60 days on May 22.

77. **The volume of water flowing through a pipe varies directly with the square of the radius of the pipe. If the water flows at a rate of 80 liters per minute through a pipe with a radius of 4 cm, at what rate would water flow through a pipe with a radius of 3 cm?**
   (Rigorous) (Competency 0015)

A) 45 liters per minute
B) 6.67 liters per minute
C) 60 liters per minute
D) 4.5 liters per minute

Answer: A

Set up the direct variation: $\dfrac{V}{r^2} = \dfrac{V}{r^2}$. Substituting gives $\dfrac{80}{16} = \dfrac{V}{9}$. Solving for V gives 45 liters per minute.

78. If three cups of concentrate are needed to make 2 gallons of fruit punch, how many cups are needed to make 5 gallons?
(Easy) (Competency 0015)

A) 6 cups
B) 7 cups
C) 7.5 cups
D) 10 cups

Answer: C

Set up the proportional equation in terms of the ratios of concentrate (C) to fruit punch (P):

$$\frac{C}{P} = \frac{3\,cups}{2\,gallons}$$

Solve for the amount of concentrate, given 5 gallons of fruit punch.

$$\frac{C}{5\,gallons} = \frac{3\,cups}{2\,gallons}$$

$$C = \frac{3\,cups}{2\,gallons}\,5\,gallons = 7.5\,cups$$

79. Which of the following best describes the process of induction?
(Average Rigor) (Competency 0016)

A) making an inference based on a set of universal laws
B) making an inference based on conjecture
C) making an inference based on a set of concrete examples
D) making an inference based on mathematical principles

Answer: C

Induction is the process of making inferences based upon specific, concrete cases or examples. Natural science, for example, uses induction to generalize results from specific observations so as to make broader statements about the characteristics of the universe.

80. Which of the following is a valid argument?
(Average Rigor) (Competency 0016)

A) Given: if p then q; q. Therefore: p.
B) Given: if p then q; ~p. Therefore: ~q.
C) Given: if p then q; p. Therefore: q.
D) Given: if p then q; ~q. Therefore: p.

Answer: C

If the premise of a given conditional statement ("if p then q") is true (that is, it is a given), then the conclusion must likewise be true. Answer C, then, is the correct answer. The other options commit various logical fallacies. These arguments may be dealt with by using specific statements in place of the symbols p and q. For instance, let p be "I am in New York City," and let q be "I am in New York." Obviously, if p is true then q must be true as well. Given that "I am in New York City," (that is, given p), then it must be true that "I am in New York" (meaning that q is true, thus demonstrating C to be a valid argument).

XAMonline, INC. 21 Orient Ave. Melrose, MA 02176

Toll Free number 800-509-4128

TO ORDER Fax 781-662-9268 OR www.XAMonline.com

GEORGIA ASSESSMENTS FOR THE CERTIFICATION OF
EDUCATORS   -GACE - 2008

PO#   Store/School:

Address 1:

Address 2 (Ship to other):

City, State Zip

Credit card number_____-_____-_____-_____   expiration_____

EMAIL _____

PHONE   **FAX**

13# ISBN 2007	TITLE	Qty	Retail	Total
978-1-58197-257-3	Basic Skills 200, 201, 202			
978-1-58197-528-4	Biology 026, 027			
978-1-58197-529-1	Science 024, 025			
978-1-58197-341-9	English 020, 021			
978-1-58197-569-7	Physics 030, 031			
978-1-58197-531-4	Art Education Sample Test 109, 110			
978-1-58197-545-1	History 034, 035			
978-1-58197-527-7	Health and Physical Education 115, 116			
978-1-58197-540-6	Chemistry 028, 029			
978-1-58197-534-5	Reading 117, 118			
978-1-58197-547-5	Media Specialist 101, 102			
978-1-58197-535-2	Middle Grades Reading 012			
978-1-58197-539-0	Middle Grades Science 014			
978-1-58197-345-7	Middle Grades Mathematics 013			
978-1-58197-546-8	Middle Grades Social Science 015			
978-158-197-573-4	Middle Grades Language Arts 011			
978-1-58197-346-4	Mathematics 022, 023			
978-1-58197-549-9	Political Science 032, 033			
978-1-58197-544-4	Paraprofessional Assessment 177			
978-1-58197-542-0	Professional Pedagogy Assessment 171, 172			
978-1-58197-259-7	Early Childhood Education 001, 002			
978-1-58197-548-2	School Counseling 103, 104			
978-1-58197-541-3	Spanish 141, 142			
978-1-58197-610-6	Special Education General Curriculum 081, 082			
978-1-58197-530-7	French Sample Test 143, 144			
			SUBTOTAL	
FOR PRODUCT PRICES GO TO WWW.XAMONLINE.COM			Ship	$8.25
			TOTAL	

Printed in the United States
141597LV00003B/25/P

9 781581 973464